Behind the Front Panel

Vintage Radio Enthusiasts

Collectors Amateurs Curators Engineers

Here's a book that gives the story of the design and development of 1920's radios. The answers of what goes on "behind the front panel":

How do all those shiny tubes and pretty components work?

Why did the old engineers use so many different circuits?

What was the role of men like Armstrong, Fleming, De Forest, Marconi, Alexanderson, Hazeltine and others?

How did RCA's patents lead to the design of new circuits?

Why did the triode dominate radio design for over 10 years?

Who invented the Neutrodyne, Superdyne, Technidyne, Isofarad, Counterphase, Syncrophase and the Superheterodyne?

When did the first "one knob" radios appear?

These and many other questions are answered in a fascinating story using photographs and simple illustrations that can be understood by anyone having an elementary knowledge of electricity.

The author's extensive research has provided many amusing and anecdotal quotations from 1920's popular and technical magazines.

40 Illustrations
25 Photographs
75 References
Glossary and Index
as well as citing 45 different manufacturers of the period.

From Crystal Sets to Eight Tube Superhets

Behind the Front Panel

The Design and Development of 1920's Radios

David Rutland

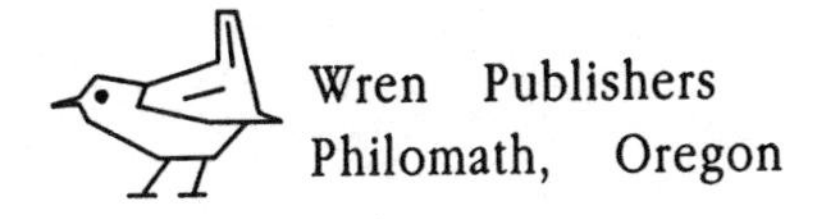

Wren Publishers
Philomath, Oregon

Wren Publishers
P.O. Box 1084
Philomath, OR 97370
(503) 929-4498

Printed in the United States of America

ISBN 1-885391-00-5
Library of Congress Catalog Card Number: 94-60507

Publisher's Cataloging in Publication Data
Rutland, David
Behind the front panel: the design and development of 1920's radios / David Rutland

x, 142 p., [13] p. of plates: ill. ; 22 cm.
Includes bibliographic references (p.[137] to 140) and index
1. Radio—Receivers and Reception—Design and Construction—History. I. Title
TK 6547.R 1994
621.384'18'09042 — dc20

To Jason and Heather

Table of Contents

List of Plates

Preface

About five years ago I happened to look through a buyers guide to antiques. I was intrigued when I came across some antique radios. Did the old radios still exist? Did someone collect them? A little more investigation revealed that indeed many antique radios do exist and there are many collectors throughout the world. I found and joined the Antique Wireless Association (AWA), the Antique Radio Club of America (ARCA) and one close to home, the North West Vintage Radio Society (NWVRS). Many of the members in these societies are interested in establishing a collection as a collection in itself. Others take an interest in restoring the radios to their original condition. Still others wish to preserve the radios for future generations and their collections are veritable museums. Each of the societies have established museums where radios dedicated to the museum by their members are displayed. Not only old radios are shown but also vacuum tubes, old posters and radio magazines.

When I was in high school in the late 30's I became interested in electronics and my father gave me some 1920's radios. Like other experimenters I promptly took them apart and used their parts for my own projects. I wish now I had kept one! But looking at all the old radios the collectors had I was reminded of my younger days and, as an electronic engineer, I began to wonder how these radios differed, not from external appearance, but from the electronics inside. So unlike some of my friends at NWVRS I look more closely at the "innards" than at the cabinet.

I quickly found that each radio had its own unique circuit and component layout and my curiosity was raised sufficiently for me to start on a research program as to the how and why of the circuits. Having worked with vacuum tubes for the first twenty years of my engineering career I was in a good position to understand the old triode tubes and the theory of their operation. My search for information finally led me to the public library where I was pleased to find good collections of the old radio magazines. The 20's is really ancient history today and many times I had to ask the librarian to get the old magazines out of storage. What I found was a multitude of descriptions and facts on the old radios. I found so much that didn't seem to be available in other books on antique radios that I decided to incorporate what I learned in this book.

I'm grateful for the librarians at the Corvallis Public Library, the Multnomah Public Library and the Oregon State University library for their help and patience in retrieving these old volumes. I am specially grateful to the members of the antique radio clubs for getting me interested in a subject that I had no idea I was ever going to be involved in. Without them I would never have been able to acquire the few old radios in my collection and would never had written this book.

D. R.
March 1994

Behind the Front Panel

Chapter 1

Radio Beginnings

The decade of the 20's saw the flourishing of broadcast radio. At the beginning there were a few radios built by amateurs and by the end of the decade there were millions of radios in use by the general public. Before the decade started major radio engineering developments had progressed to the point where it was possible to build radios for public use. The pioneering inventors J. A. Fleming and Lee De Forest had produced the radio tube. The engineers Armstrong and Hazeltine had developed practical circuits using the new tube. As early as 1913 the vacuum tube was being used by the telephone company to make long distance telephone conversations a reality. These developments and many others laid the groundwork for the radio receivers of the 20's.

Transmitters for the sending of voice and music were also developed before 1920. The first radio transmitters used high voltage spark generators and were only suited for sending messages by telegraphic code. These spark transmitters sent out waves covering a wide frequency band causing nearby transmitters to interfere with each other. This interference was solved by the invention of transmitters that generated continuous waves on a single frequency. Before vacuum tubes continuous waves were produced by two types of radio frequency generators, the Poulsen arc and Alexanderson's high frequency alternator.

Poulsen arcs were used extensively by De Forest for experimental broadcasts. These machines used a continuous electric arc operating in a strong magnetic field. The larger Poulsen arcs had electromagnets taller than a man. Even though smaller units were built these transmitters were generally used by commercial communication stations rather than by amateurs. Alexanderson alternators were also primarily used for commercial communications. These were rotating machines driven from electric motors or steam turbines like those that generate electric power. Practical radio requires currents having a much higher frequency than the 60 Hz used by electric power companies. Frequencies as high as 100,000 Hz were generated by the special alternators that Alexanderson designed. His machines rotated at high speeds, about 20,000 RPM, and had hundreds of magnetic poles and generated considerable power. The low radio frequencies required giant antennas covering many acres and gave reliable telegraphic communication across the Atlantic. Even as late as December, 1921, President Harding was at the opening of what Radio News described as the "World's Most Powerful Station". It was planned to use as many as 10 Alexanderson alternators each capable of generating 200 kilowatts of power and transmitting at the very long wavelength of 16,400 meters or 18,000 Hz.

The man who first sent a radio signal across the Atlantic, Guglielmo Marconi, was a steadfast believer in these very low frequency radio signals and never stopped using them. But others started experimenting with "short" waves. These "short" waves were not short at all by modern standards but were the medium wavelengths now used in the AM broadcast band, 550 to 200 meters (550 to 1500 kHz). The words "shortwave" and "longwave" have acquired different meanings during the history of radio which can result in some confusion when reading the old literature. The "short" waves of the '20's are now "medium" waves and the "long" waves are now "very long" waves.

Although it was possible to use the Alexanderson alternator and Poulsen arc to transmit voice, the wide use of voice transmission had to await the less expensive vacuum tube. The transmission of voice and music then became practical and in 1920 and 1921 the popular radio magazines were full of construction articles for "radiophones", what we would now call transceivers or walkie-talkies. The radiophones transmitted "short" waves and therefore they could have been heard on today's AM radios. Many were truly portable, the size of a small suitcase, and provided the radio amateurs an affordable means to communicate with each other. Some amateurs sent music over their radiophones for the entertainment of their friends and families. They possibly could be called the first broadcasters. Partly because of these amateur broadcasts many people in radio, including David Sarnoff of RCA, were talking about broadcasting for the general public. Like the chicken and the egg, there could be no broadcasts without broadcast stations and no broadcast stations unless the public had radio receivers. The radio stations had to come first on the gamble that the public could be convinced that having a radio would be worthwhile and be entertaining. Westinghouse is credited to be the first company to take this gamble and on November 5, 1920 their famous station, KDKA, went on the air with regularly scheduled evening programs.

The pace then began to pick up. A glimpse of the quickening of this pace is provided by a glance through the popular magazine Radio News. In June 1921 they announced that DeForest, who had his radio company in California, had installed a transmitter at the California Theater in Los Angeles. They reported that the California Theater Radio Station had been transmitting concerts over the air for over a year. That's before KDKA. Were they the first broadcaster? It depends on how one defines broadcasting. Is it broadcasting when you do it now and then like the hams, or at weekly or monthly concerts, or every evening like KDKA or day and night like today's stations?

Then there's the report that in October 1921 a certain Charles L Austin in Portland, Oregon, was using a transmitter to advertise phonograph records from his record store. Each record selection was preceded by the title and the name of the store. Was this the first commercial? Or was he the first disc-jockey? Probably not, it is said that some of the early amateurs were able to profit from the announcement of a sale in a local store. In either case, the seeds of commercial radio had been laid.

In August 1921 the Missouri State Agriculture Department was planning to buy a transmitter to broadcast farm news and commodity prices. In November a large electrical show was held in New York City. The reports by Radio News commented that "Concerts by radio are now common" and "If in each home a receiver with a loudspeaker were installed it would replace the phonograph". Note that radios required "installation" and were not just taken home and turned on. Not only must the purchaser supply a good antenna and ground but he would have to hook up all the batteries and learn how to operate all the controls.

Farmers were receiving reports on an experimental basis by November. Back in September E. Tunney announced a new receiver that was specifically designed to receive only broadcast band wavelengths and thereby catered directly to the broadcast listener. In December that year kits for simple radios appeared in the department stores for the first time.

In 1922 things really started to hum. In January it's reported that "Radio business is good" and "the demand for radio apparatus is higher than ever". Radio apparatus had meant radio parts for building your own receiver but now a half dozen receiving sets were on the market. Advertisements for complete radios appeared—Clapp-Eastham in January, Grebe in February, Tuska in June and Crosley in September. Quite a change from the year before when there were only radio parts advertised.

The Chicago Grand Opera was broadcast over Westinghouse's fourth transmitter, KYW, in February, the United States Post Office broadcasted weather and market reports from Washington, D.C. starting April 15, General Electric's WGY at Schenectady started intermittent operation, and the Signal Corps WYCB entertained the troops in the evenings. Church services were broadcast at Easter. The Bureau of Standards was swamped by letters requesting their pamphlet on how to build a simple radio receiving set. The Brooklyn Radio Show exhibited complete receivers.

Up to now, all the broadcast transmitters were assigned one wavelength, 300 meters (1000 kHz), relying on distance separation to prevent interference. With so many transmitters this single frequency immediately became crowded so that new wavelengths were assigned on May 15, 1923. By the end of that year literally hundreds of broadcast stations were on the air and the 20's radio boom was in full swing.

So what electronic circuits did these new commercial home "radio receiving sets" use? How did the circuits change during the twenties? What drove the designers to dream up new circuits? And why was it that for 10 years from 1917 to 1927 the simple triode tube was the only type used in these sets?

The purpose of this book is to throw some light on these questions. The history of the radio manufacturing companies themselves has been more than adequately covered elsewhere (Douglas 1988). Here the emphasis will be placed on the electronic circuits that were used in these pioneering radios. These circuits will be described and, whenever possible, credits given to the engineers who designed them and the companies that manufactured them. The motives driving these engineers to invent the new ideas were much the same as they are today. Many owned the companies that produced the radios and were swayed by the economic pressures of the marketplace. Others found satisfaction in the design of new and better circuits. Many of those who put in long hours of experimentation have not left a record and must remain anonymous. In those days, engineers, even if they did have a college education, were facing a brand new field of engineering and had to learn the ropes through experience. Professors at the universities were kept busy trying to develop the theory behind the operation of their circuits. But many professors in the electrical engineering and technology schools also contributed to the practical advancement of radio receiver design. These practical no-nonsense engineers knew what worked and what didn't and, at the behest of their company owners, produced a wide variety of receiver circuits.

Some of these circuits would strike modern engineers as being without merit. But if the old-time engineers could talk today they would, no doubt, be able to give good reasons for their designs. Many of these good reasons must be inferred from the study of the designs themselves. Their circuits didn't necessarily come from the desire to produce the highest performance or to incorporate the latest ideas. But they were guided, as many engineers still are, by the need to produce sets of low cost, easy to use by the layman, and at the same time avoid the payment of high royalties to patent holders. Their circuits must also have had a different "angle " that would appeal to the trade magazines, got their company a write up in the monthly radio magazines, and provided a reason for the salesman to push their sets.

In those days a sophisticated radio buyer would care as much about the internal workings of their sets as they would about the aesthetics of the radio cabinet. Radios of the 20's were limited in performance and so they were chosen by customers on their technical merits. Today it goes without question that the smallest "transistor" has adequate sensitivity and selectivity. All radios have a simple one knob or automatic tuning so that the desired station can be immediately selected with little effort. Although this state of the art wasn't reached until the 30's, the customer's requirements were much the same and the designers of the 20's radios strove to meet them with the modest technology at their disposal.

Descriptions of electronic circuits are difficult to give without circuit diagrams and the descriptions in this book are no exception. In order to help the reader who may not be acquainted with electronic schematics the Appendix provides a list of symbols and a review of technical terms both old and new. Definitions of technical words are provided in the Glossary.

The schematics shown in this book are "simplified". That is they emphasize the flow of signals through the circuit and neglect other aspects of practical circuits. For example, the batteries and circuits that provide power to the vacuum tube filaments are not shown. The function of these circuits is to provide the current to heat the filaments to operating temperature. They are necessary to make the tube operate but otherwise do not enter into the description of the tube's important amplification function. The omission of these circuits is indicated by the lack of a connection to one end of each filament.

Another degree of clarity is obtained by showing separate batteries for each tube whereas, in a radio with more than one tube, a practical circuit would combine these into one single battery.

Each of the following chapters treat the development of one type of circuit: detectors, amplifiers, superheterodynes and so on. They are arranged in order from the more basic circuits of early radio up to the complete receivers of the late 20's going from the simple to the more complex.

Chapter 2

Detectors I From Sparks to Crystals

When Isaac Newton was congratulated for his great achievement, the Theory of Gravity, he said that he had only stood on the shoulders of giants. When James Clerk Maxwell wrote down his four equations that predicted electromagnetic radiation (radio waves) he also was standing on the shoulders of giants. These giants are known to all who study or experiment with electronics as the names given to the units of electrical measurement Georg Ohm, Alessandro Volta, André-Marie Ampère, Joseph Henry, Michael Faraday and Heinrich Hertz. It was Hertz who first confirmed Maxwell's theory by generating and detecting radio waves. His "transmitter" used a high voltage spark across the gap of a loop antenna. This he placed at one end of his laboratory and at the other end he placed a similar loop with a narrow spark gap. When he energized the spark in his transmitter it produced a high frequency alternating electric current in the loop. These radio frequency (RF) currents produced electromagnetic radiations or "radio waves" which were transmitted across the laboratory to the receiving loop. The induced voltage in this loop was

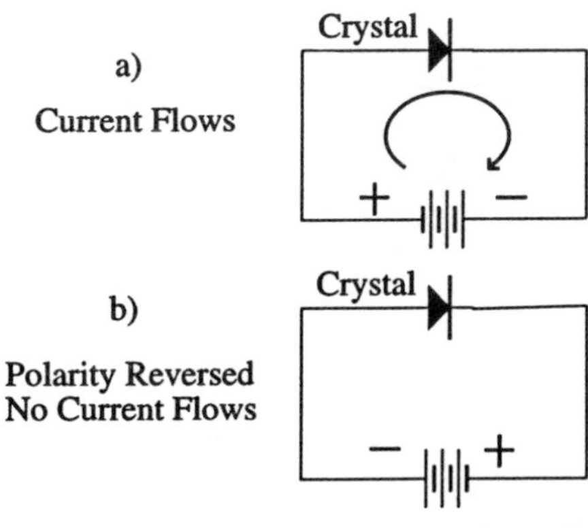

Fig. 2.1 Crystal

sufficient to produce a tiny spark across the gap and he had become the first man to detect radio waves.

Although it might be said that he detected the radio waves using a spark, actually the spark only indicated the presence of the radio frequency currents induced in the loop by the waves. And so it has been in all radios since: the presence of radio waves is indicated by using an antenna and a detector of RF currents or voltages.

The spark "detector" used by Hertz was much too insensitive for the reception of radio waves over distances outside the laboratory. Much more sensitive detectors were soon invented to make practical radio communication possible. Guglielmo Marconi used Morse code to send signals by turning his transmitter on in short bursts to represent dots and dashes. He was first to show that radio was practical by using a more sensitive detector, the coherer, a mixture of iron filings that cohered or stuck together when the RF current from his antenna ran through them. Normally when the iron filings were loose they had a high resistance which was lowered when they cohered. This change in resistance provided the means to operate an electromagnet whenever a radio signal was received. The magnet operated a metal arm that tapped the coherer and un-

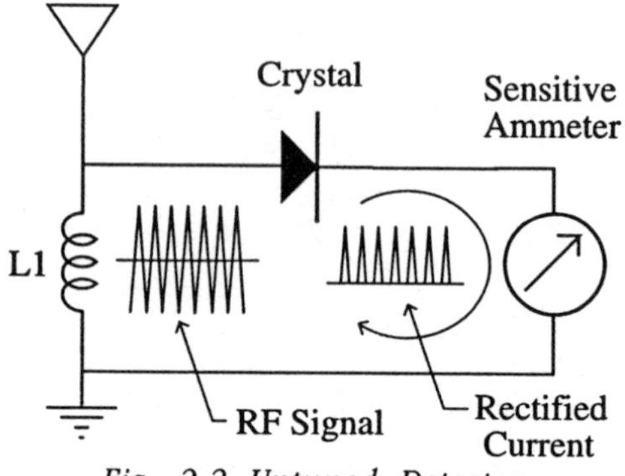

Fig. 2.2 Untuned Detector

stuck the cohered filings. The operator heard the taps and was able to read the telegraphic signal.

Later on, certain crystals were found to have detecting properties when a fine wire, the "cat-whisker", was lightly pressed against their surface. These crystals operated as a detector because they responded differently to a positive than to negative voltage. As shown in Fig. 2.1, applying a voltage of the proper polarity across a crystal will readily cause a current to flow. However, when the polarity is reversed only a small current will flow.

The crystal's rectifying action may be used to detect radio waves as shown in the simple detector circuit, Fig 2.2. The alternating RF signal from the antenna appears across the inductor, L1, connected between the antenna and ground. A crystal and a sensitive ammeter (galvanometer) completes the circuit. Only the positive peaks of the RF signal pass through the crystal. The graph on the left of the figure shows the RF voltage across the coil alternating from positive to negative and the right hand graph shows the rectified unidirectional signal. The average value of the rectified peaks is indicated by the ammeter. The presence of a radio signal has been detected.

It is interesting to note that although the old radio engineers knew of the crystal's rectifying action they did not know

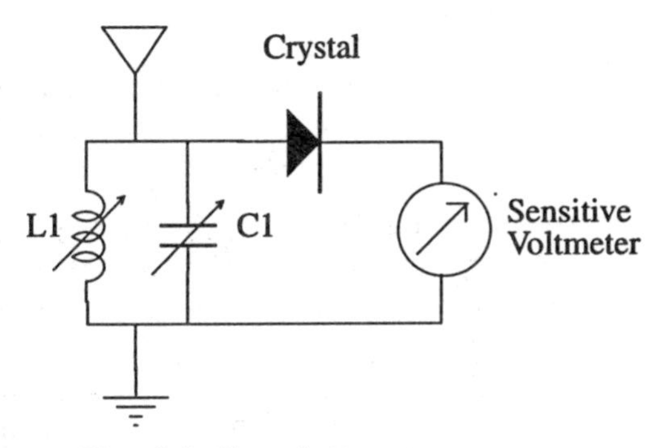

Fig. 2.3 Tuned Detector

the physical basis of its operation. It wasn't until after World War II that solid-state atomic physics was understood sufficiently to explain its operation. This understanding led Schockley, Bardeen and Brittain to invent the transistor. In fact, the first transistors weren't all that different from the old "cat-whisker" crystal as they actually used two "cat-whiskers"!

At the turn of the century Marconi was first attempting his long distance transmissions using his coherer. Like the simple crystal detector, Fig. 2.2, the coherer was connected directly from his antenna to ground. Any radio signal of sufficient strength, regardless of its frequency, would be detected. At first, the only signal was Marconi's but, later, as more transmitters came into use, two nearby transmitters would be heard simultaneously. This interference produced an amusing incident at the International Yacht Races held near New York in 1901. Both Marconi and DeForest had contracted with rival newspapers to report the progress of the race by radiotelegraph. They each had a transmitter in a tug boat near the race and a receiving station on shore. When the race began they both kept up sending continuous messages to shore. The interference was so bad that no messages could be received by either shore station. In spite of this the two New York newspapers each claimed that their reports had been received by radio! (Archer 1938)

At the time of the yacht race, the English scientist, Sir Oliver Lodge, had already found a solution to this interference problem. In 1897 he had experimented with an ingenious apparatus that would send out radio waves of a definite frequency. He devised receiving sets using electrically tuned circuits so that only waves of the desired frequency were received. The simple crystal detector, Fig. 2.2, receives waves of all frequencies. It can be modified, Fig. 2.3, to receive waves of a definite frequency by adding the capacitor, C1, across the inductor, L1. The inductor and capacitor, L1 and C1, form a "tuned" circuit that responds to a small band of frequencies and rejects others. The exact frequency to which it responds is called the "resonant" frequency and is determined by adjusting the values of L1 and C1. The small band of frequencies above or below the resonant frequency which are not completely rejected is called the "bandwidth" and can be made wide or narrow by the proper design of the circuit. The bandwidth depends primarily on the resistance inherent in the windings of the inductor. When the resistance is low the bandwidth is narrow and the circuit is said to have high "selectivity". The capacitor, C1, is usually made by placing two or more plates closely parallel and insulated from each other. When properly constructed these capacitors introduce very little resistance in the tuned circuit.

DeForest was a pioneer in radio broadcasting and was one of the first to transmit voice and music over radio waves. The system he used we now call amplitude-modulation or "AM". Early communication systems using voice were called "radiotelephones" because they transmitted telephone signals over the radio. The telephone transmits the sound waves from the human voice by using a microphone to convert them to an electrical signal. These converted sound wave signals have a much lower frequency than radio waves. When these "audio frequency" (AF) signals are applied to a radio transmitter so as to vary the strength of its transmitted signal the RF signal so produced is said to be "amplitude modulated". The transmitted radio waves then rapidly vary in strength at the audio frequency rate. At the receiving end these amplitude modulated (AM) waves must be converted back to audio frequencies to recover the original signal.

The principles of AM transmission are illustrated in Fig. 2.4. The RF output of the transmitter, the "carrier" wave, (b), is varied in strength by the AF signal, (a). Negative peaks of the AF signal decrease the amplitude of the carrier, positive peaks increase the amplitude. The result is the amplitude modulated wave, (c), which is transmitted to the receiver. The receiver uses a detector which, like the crystal, rectifies the AM signal and produces an unidirectional current, (d). This current is not steady but its strength or amplitude varies in the same manner

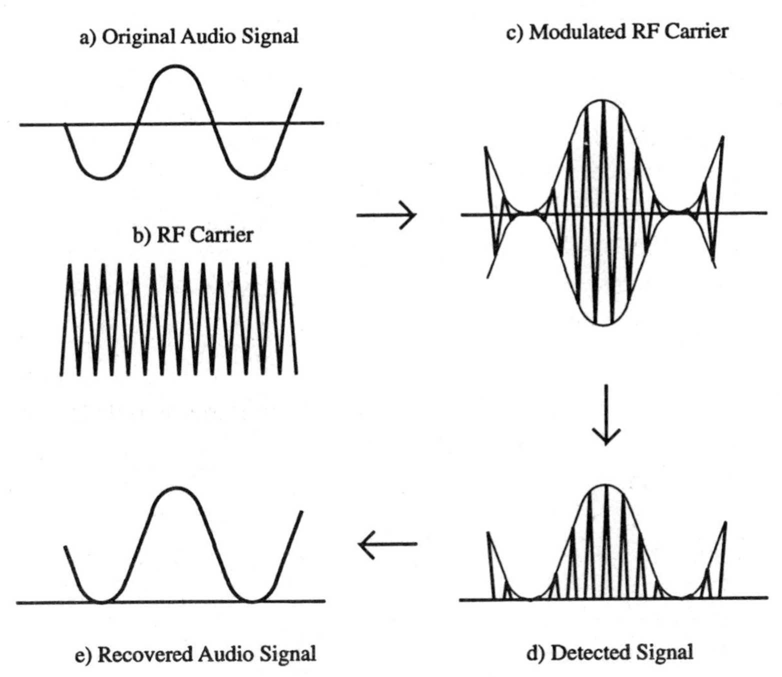

Fig. 2.4 Amplitude Modulation

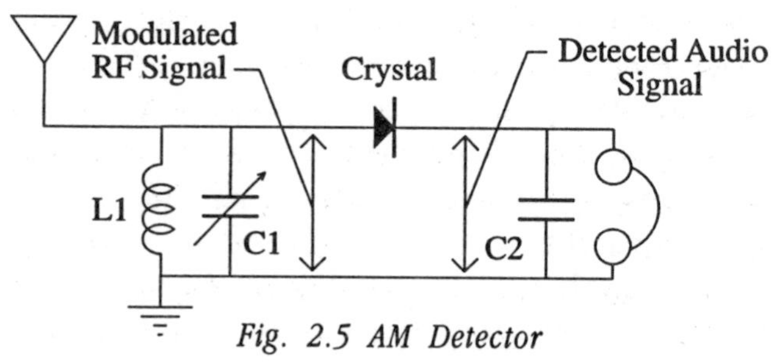

Fig. 2.5 AM Detector

as the original AF signal. The average value of the detected signal, (e), reproduces of the original AF signal and the transmitted sound signal has been recovered at the receiver.

The simple detector circuit, Fig. 2.3, can be used to receive AM waves by simply replacing the meter with a set of sensitive headphones, Fig. 2.5. With this crystal radio the AF signal transmitted on the AM RF signal could then be heard. The broadcasting of speech and music over the "air" had become a reality.

Before vacuum tube amplifiers were developed for transmitters the microphone directly controlled the RF carrier by being placed between the transmitter and antenna. Most stations generate thousands of watts of energy which had to pass through the microphone. In order to handle this power multiple microphones were not uncommon and many microphones had to be water cooled!

The invention of AM gave a different meaning to the word "detector". At first it meant a device to indicate an RF electric current from an antenna but now it has changed to mean a device to recover the AF signal from an AM signal, that is, a demodulator. In fact, "detector" continues to refer primarily to circuits used to recover radio signals even when no amplitude modulation occurs. For example, a modern FM (Frequency Modulation) radio has an FM "detector" that detects only changes in frequency.

The simple crystal detector, Fig. 2.5, using a single inductor or coil and a variable tuning capacitor, can be made to provide better selectivity by using two coils electromagnetically coupled to each other. An example of the use of two coupled coils in an early crystal radio is shown in Fig. 2.6. The two coils, known as an "antenna coupler", were wound using a single layer of wire on hollow plastic cylinders. The smaller diameter coil, L1, was arranged so that the listener can slide it in and out of the larger coil, L2. As the coils were withdrawn less of the signal from the antenna was induced or coupled to the secondary, L2, but the selectivity was increased. So that when the coupling was made small faint signals were difficult to receive while when the coupling was made large the selectivity of the circuit was impaired. The listener had to slide the coils in and out to choose a compromise between these two conditions.

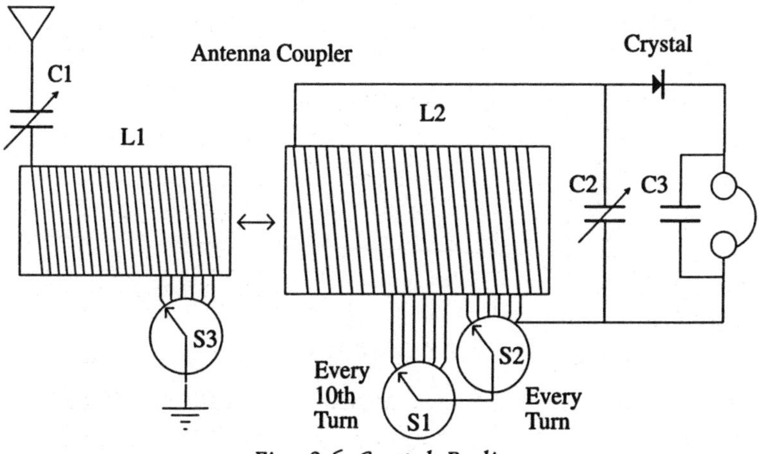

Fig. 2.6 Crystal Radio

The induced RF signal voltage across L2 was then detected by the crystal and heard on the headphones. The capacitor, C3, provides a low impedance path for the radio frequency signals and allows only the audio frequency signal to go through the headphones.

The two coils in the crystal radio were tuned to the frequency of the desired station by the capacitors, C1 and C2. The schematic shows that the tuning was accomplished by adjusting both the values of the two coils, L1 and L2, as well as the capacitors, C1 and C2. Many early sets fixed the value of the capacitors and used variable coils while later sets used fixed coils and used variable capacitors.

The crystal set coils were "tapped", that is, provided with many connections along their lengths. The inductance was varied by switching to different taps thereby inserting or removing turns from the coil (Plate 3). The schematic shows two different methods. The antenna coil, L1, is shown using a single rotary multi-position switch, S3, to select the different taps while the secondary coil, L2, is shown using two switches. One switch, S1, was for coarse adjustment of the tuning and selected taps spaced ten turns apart while the other, S2, was for fine adjustment and selected taps placed only one turn apart. At times even the step from one turn to another was too coarse to tune in exactly to the station frequency so that the variable capacitors had to be used for final tuning.

By 1920 continuous control of inductance was obtained by splitting a coil in two parts and rotating one within the other. The rotating part of the coil, the rotor, was made smaller so that it could be freely rotated inside the larger part, the stator. The two parts were connected in series so that when the rotor was aligned with the axis of the stator the mutual coupling would add to the total inductance. As the moving coil was rotated the coupling would gradually decrease, decreasing the effective inductance of the two coils. When the coil had been rotated a half turn the coupling would subtract inductance from the two coils and the total inductance would be at a minimum. In order to obtain a greater change in inductance the coupling was increased by winding the coils on spherical forms, one smaller than the other. The smaller spherical coil was mounted on a shaft inside the large coil so that it could be provided with a knob and rotated by the listener. This device was widely used in the 20's and was known as a "variometer" (Plates 1 and 2).

The variometer went out of use with the development of variable capacitors which used semi-circular movable plates rotating close to similar fixed plates. Large variable or tuning capacitors using multiple plates became the preferred choice of most radio designers in the 20's (Plate 8). However the use of variable inductors was not completely abolished until the 30's.

The sliding concentric coils of the early crystal set that were used in the antenna coupler were soon replaced by the "variocoupler" which, like the variometer, rotated one coil within the other (Plate 3). The inner coil was made short enough to be able to rotate, rather than slide, within the outer, larger coil. Unlike the variometer, the variocoupler coils were wound on cylindrical rather than spherical forms. But like the variometer the rotating coil was mounted on a shaft and controlled by a knob on the front panel of the set.

Although, to get good reception, it took skill to adjust the coil, capacitor and crystal of these old crystal sets the circuit was simple and, even when tube radios became available, it was more than likely to have been the first radio that a broadcast listener would have used. It did not require batteries and with simple instructions it was relatively easy to build. Literally millions of these "crystal sets" were produced over the years and are still available in kits for students and hobbyists. Critical and patient placing of the "cat's whisker" against the crystal's surface would eventually yield good results. Today, everyone who is familiar with solid-state electronics, transistors and integrated circuit chips knows the crystal as a simple solid-state diode. Modern constructors of crystal radios may try to use the old crystal with a wire contact but, when the novelty wears off, they can easily get good reception by substituting a commercial solid-state diode.

Chapter 3

Detectors II From Crystals to Tubes

It was the search for a reliable detector that prompted John Fleming, a technical advisor to Marconi in England, to invent his "oscillation valve", the two-element or diode tube. His invention was based on the "Edison effect" which was discovered when the inventor of the electric lamp noticed an emission from the lamp's filament that darkened the glass envelope. This emission was later proved to be caused by the emission of electrons from the filament but Edison's work preceded the discovery of electrons by many decades.

In 1906, Fleming added a metal plate, or "wing" as it was then called, to an ordinary lamp bulb and connected it to a separate wire which he brought out through the bulb, Fig. 3.1. The lamp's filament was heated white hot by current from the "A" battery so that it would emit electrons. When the "B" battery was connected as shown to make the plate positive with respect to the filament it attracted the negatively charged electrons. An electric current then flowed continuously through the "B" battery back to the filament. If the "B" battery was reversed, the plate was made negative and repelled the electrons so that no

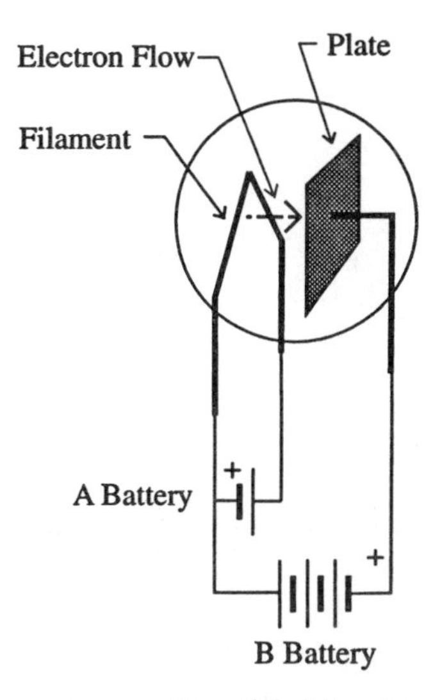

Fig. 3.1 The Diode

current flowed. Thus the two elements, the filament and plate, exhibited the same rectifying action as the crystal.

Fleming already knew that Edison had discovered this rectifying action and wanted to know if it would operate as a detector. Fleming therefore tried his tube in a circuit similar to the tuned crystal radio, Fig. 2.3, replacing the crystal with the two-element tube. He found that it did indeed detect radio frequency currents and he obtained a response on his sensitive ammeter. It is said that Fleming was hard of hearing and couldn't hear the clicks generated by a coherer so he was very pleased to have found a detector with a visual indicator. He called his tube an "oscillation valve" and resisted using the modern name "diode" meaning "two paths".

When Fleming invented his diode detector he, like Edison before him, did not know about electrons although scientists had recently discovered them. Many thought that the operation was caused by ionized gas which wasn't far from wrong as his tube, although evacuated, still contained residual gasses. The

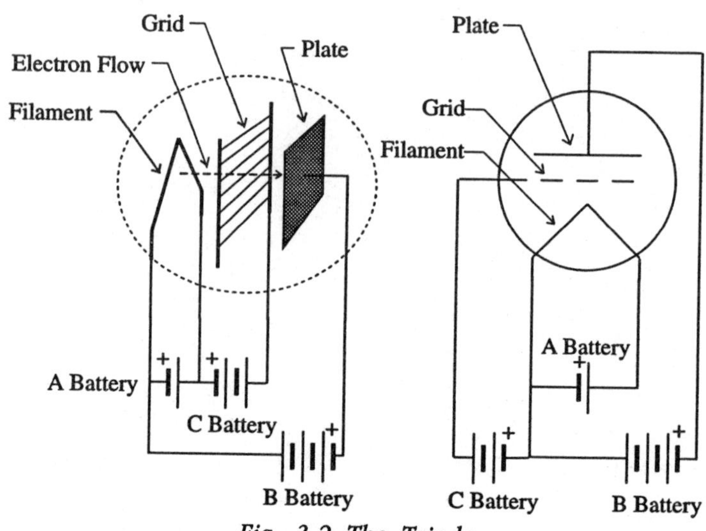

Fig. 3.2 The Triode

gas indeed had a small influence on the diode's operation but the primary effect was caused by electrons from the filament. All early tubes were made by electric lamp manufacturers who didn't have the high vacuum pumps later used for evacuating tubes. In hindsight it was perhaps fortunate that high vacuum techniques were not available to the lamp industry as the electric lamp filament's life is extended by traces of gas inside the envelope. Thus, although Fleming invented the tube detector he did not invent the modern vacuum tube diode. The true vacuum tube was not developed until 1913 by Arnold of Western Electric (Chapter 13).

Lee De Forest is known as the "inventor of the vacuum tube" but he also did not know about the effect of residual gas when he invented his "audion" tube in 1906. Fleming's diode was not a very sensitive detector and De Forest was looking for ways to make a better detector. He was experimenting with tubes with more than two electrodes, trying different arrangements. After much trial and error he placed a mesh between the filament and the plate, Fig. 3.2. The mesh was shaped like a gridiron and was henceforward called the "grid". His audion thus contained three electrodes and now goes by the modern name "triode".

Like Flemings diode, electrons emitted from the triode's filament are attracted to the plate that's held positive by the "B" battery. The grid is made negative by the "C" battery and repels some of the electrons preventing them from going through to the plate. If the grid is made more negative by increasing the C battery voltage more electrons are repelled and the electron flow is reduced and less current flows to the plate. The schematic symbol for the triode is shown on the right, Fig. 3.2, connected in the same way as shown on the left.

De Forest's audion was constructed as shown in the sketch with a flat grid and plate. Later tubes were designed with a more efficient configuration by wrapping the grid and plate around all sides of the filament. In either configuration the operating principles are the same.

The grid's control of the electron stream made the triode a sensitive detector. The detector circuit, Fig. 3.3, like the crystal detector, Fig. 2.5, uses a tuned antenna circuit, L1 and C1. The antenna is connected to a tap on L1 so that a larger RF signal voltage is developed across the tuned circuit. The signal is connected to the filament and grid through the resistor, R2, in parallel with the capacitor, C2. The plate is supplied by the "B" battery through the headphones.

This simplified schematic does not show the filament "A" battery and rheostat which provide the current for the filament. Of course these components must be present to heat the filament so it will emit electrons otherwise the tube would not work. However they are not necessary for the description of the tube's primary operation and are omitted for clarity in all schematics in this book unless the filament circuit is itself a topic of interest. The detector circuit operated the tube with little or no voltage or bias between the grid and filament. The electrons flowing past the grid to the plate were not completely repelled by the grid and some were attracted to it. This small grid or "leakage" current charged C2 until enough voltage was supplied across R2 to provide a path for the current. The resistor, R2, was called the "grid leak" as it drained off the leakage current from the grid.

Under these conditions the grid and filament performed like Fleming's diode, the grid taking the role of the diode's plate. Like the diode and crystal detectors the incoming signal was rectified and the original AF signal appeared across the grid leak, R2 and the capacitor, C2. The voltage between grid and filament now varied according to the AF voltage across R2 which, in turn, varied the plate current and produced an amplified AF signal across the headphones. Capacitor, C3, removed any RF currents that are amplified with the AF signal.

The gas in the audion made the grid have a greater effect when the plate and filament voltages were critically adjusted. Even after high vacuum tubes were developed tubes containing a controlled amount of gas were still made for detectors and used in some radios in the early 20's.

De Forest's patented his audion on February 18, 1908 and constructed transmitters and receivers for the U.S. Navy from 1907 to 1908. Over twenty Navy ships were equipped with his detector circuit when the fleet went on a pioneering world cruise.

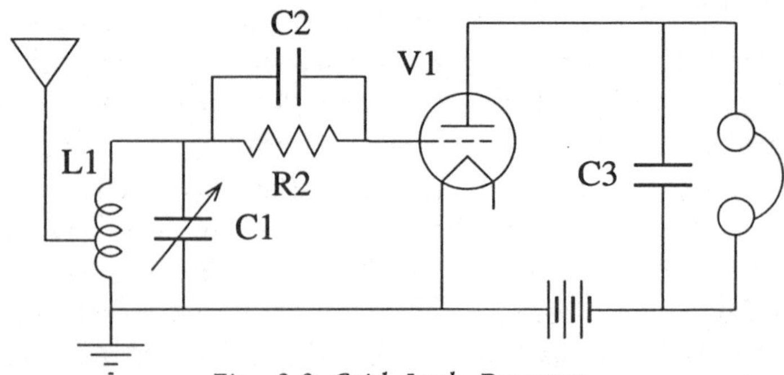

Fig. 3.3 Grid Leak Detector

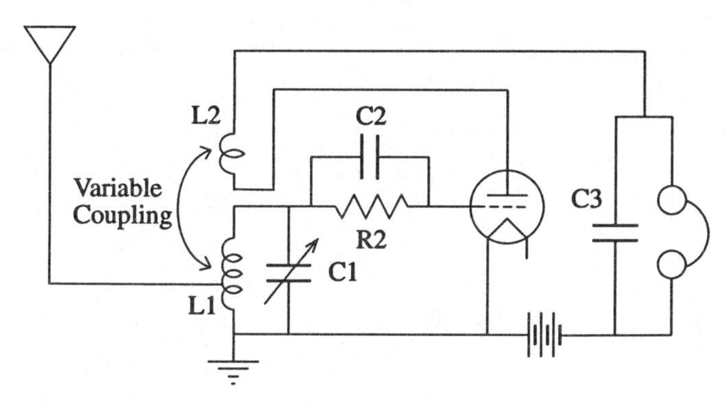

Fig. 3.4 Regenerative Detector

Even though De Forest's grid leak detector was more sensitive and reliable than other contemporary detectors its performance was greatly improved by the famous inventor Edwin Armstrong. His circuit, Fig. 3.4, was patented in 1914 and was based on De Forest's detector. Armstrong added the inductor or coil, L2, in the plate circuit of the detector in series with the headphones, and placed this coil close to the tuned antenna circuit. RF currents in the plate circuit, amplified by the action of the grid, were then electromagnetically coupled from L2 to the antenna coil, L1, so as to augment the input signal. In radio slang, the coil, L2, "tickles" the antenna coil with the amplified signal and it became referred to as the "tickler" coil. Like a variocoupler, the tickler coil, L2, was mounted on a shaft so that it could be rotated inside of the antenna coil, L1, thereby varying the amount of signal transferred from the plate to the grid circuit.

When Armstrong gradually increased the coupling by rotating the tickler, L2, he reached a critical point where enough signal was coupled back to the input circuit to produce self-sustained continuous radio frequency oscillations. In his search for a better detector he had by chance invented, and later patented, the vacuum tube oscillator, a generator of radio frequency electric currents. When oscillating, the circuit would remain amplifying its own signal, "chasing its own tail", even when the original input signal was removed. In fact the oscillations would start themselves simply from the normal fluctuations in the tube's electron flow without any stimulus from an antenna signal. With this discovery the detector circuit had become a simple transmitter. The oscillations were coupled in the reverse direction up to the antenna and radiated into space. Armstrong had invented the vacuum tube radio transmitter.

But at that time Armstrong was not as much interested in generating radio waves as detecting them. When the coupling was reduced just below the critical point of oscillation the input signal was greatly augmented by the amplified signal coupled back by the tickler coil. With the adjustment close to oscillation the simple grid leak detector had been converted to a very sensitive receiver. The input signal was really "regenerated" and the detector had such great sensitivity that it really astonished Armstrong. Thus not only had Armstrong discovered the vacuum tube oscillator but also his famous "regenerative detector".

Today, the regenerative detector operation would be described using the modern electronic engineering concept of "positive feedback". As the mathematical analysis of feedback circuits shows, positive feedback decreases the effective resistance of the input circuits and, when the coupling or feedback is large enough, the resistance can be made practically zero providing high gain. When the coupling is increased further the equivalent resistance becomes negative causing the self-oscillation.

The self-oscillating feature of Armstrong's detector was used for the reception of telegraphic CW waves. The techniques used in this application had application in later broadcast receivers. CW signals were produced by turning a transmitter on and off with the telegrapher's key. Carrying no modulation within themselves the dots and dashes were, and still are, heard on an AM receiver as a series of soft clicks, if heard at all. A special technique must be used to make them audible.

An early article written for the experimenter who wishes to build his own receiver describes receivers for CW signals and notes that "for the reception of these signals the vacuum tube is almost indispensable."(Clement 1920). The tube was "indispensable" because it, unlike the crystal, could provide the oscillations that Armstrong discovered in his regenerative detector. Clement describes the "autodyne" circuit wherein the detector tube was made to oscillate very weakly at a slightly higher or lower frequency than the incoming CW signal. This self-generated signal beats or "heterodynes" with the received signal to produce an auditory note. The beating is similar to that used by a piano tuner when he compares the frequency of his tuning fork with that of the piano string. As the piano tuner adjusts the frequency of the piano string, the frequency or pitch of the beat note rises and falls. So too as the receiver operator adjusted the tuning of the oscillating detector, the frequency of the autodyne heterodyne signal would rise and fall. The tuning was adjusted to obtain a satisfactory audio frequency so that the telegraph dots and dashes were now audible as long and short tones.

Another way to use the heterodyne principle for the reception of CW signals was to use a separate vacuum tube oscillator in place of the oscillating detector. The oscillator provided the RF signal that heterodyned with the incoming signal and was called the "beat-frequency oscillator" (BFO). The BFO has since been used on all receivers designed for CW reception. It has also had many applications in modern radio and electronic circuits. Armstrong used the heterodyne principle in a different way when he invented his superheterodyne (Chapter 14).

Broadcast radio designers found the regenerative circuit to be superior to all others and it became the first really practical circuit for broadcast receivers in the early 20's. Armstrong obtained his patent in 1914. At that time there were few

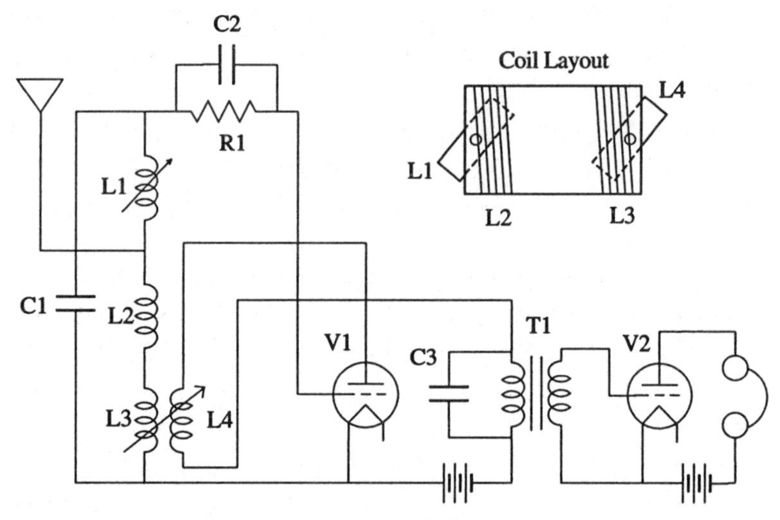

Fig. 3.5 Radiola III

commercial users so Armstrong didn't bother to charge royalties for its use. The A. H. Grebe Company was a pioneering manufacturer that used Armstrong's circuit royalty-free in their model AGP 101 advertised in QST in November 1916.

Then, during the war, everyone including Armstrong was working for the military or civilian war effort and so it wasn't until about 1920 that Armstrong started to license manufacturers. His decision was in part caused by the mounting lawyer costs in his litigation with De Forest who also claimed to have invented regeneration. One of the first commercial companies to obtain a license was Clapp-Eastham and they came out with their model HR in December 1921. Adams-Morgan, who had been making amateur regenerative radios since 1915, brought out their Model RA-10 in October 1920 which was a tuner only. That is, it contained the necessary coils but no tubes. A companion unit containing the detector tube and grid-leak was introduced in January 1921. The tuner included a tapped antenna variometer, an antenna tuning capacitor, and a variocoupler for regeneration.

Later RCA bought the Armstrong patent and subcontracted the construction of radio receivers to Westinghouse and General Electric. Their designs generally lagged behind their smaller competitors causing them to be still making sets with regenerative detectors as late as 1924. Their Radiola III illustrates the use of variometers and variocouplers rather than variable condensers for tuning and regeneration control.

The Radiola III, Fig. 3.5, uses two tubes, the detector, V1, and an audio frequency amplifier, V2, which further amplifies the AF frequency signal from the detector. The detector circuit is electrically equivalent to Armstrong's regenerative circuit. The antenna coil, L1, of the regenerative detector, Fig. 3.4, is replaced by the series combination of coils L1, L2 and L3. The variable capacitor, C1, is now fixed and tuning is accomplished by varying L1. The tickler coil, L4, couples to L3 to provide regeneration. As shown in the insert the coils L2 and L3 were wound on the two ends of a single plastic cylinder. The coils, L1 and L4, were wound on two smaller cylinders that were placed on shafts at each end of the larger cylinder. Thus L1 and L2 formed a variometer and L3 and L4 formed a variocoupler. This complete coil configuration (see Plate 4) consisted of just two fixed coils and two rotating coils and was inexpensive to manufacture. It was only necessary to add the grid leak, the AF transformer and the tubes and all the major parts for the set were assembled.

From 1917 to 1918 the United States was at war and the Navy would have liked to use Armstrong's regenerative detector but it did have one great drawback which could well have endangered the of safety of Navy ships. While an operator is using the regenerative set he tries to adjust the tickler coil for the maximum sensitivity which occurs right at the point of oscillation. If he goes past this point, which he must do to find the optimum setting, the circuit oscillates and, as we have seen, will broadcast a signal through the antenna. The enemy listening to the same frequency might receive this signal and locate the receiver's location with a direction finder. Thus even when the ship's radio transmitter was kept quiet its secret position might well have been compromised.

But it wasn't only the Navy that became concerned with this radiation from the regenerative detector. As soon as the radios became popular, people everywhere were adjusting their sets and sending out the unintended radiation. Now it wasn't the "enemy" that was listening but the neighbor next door. The radiation would produce chirps and whistles in his set and, worst of all, these same chirps and whistles would appear in the listener's set from the neighbor's radio. This state of affairs, to say the least, was not conducive to friendly neighbor-neighbor relations. Today the Federal Communications Commission has strict rules on how much any receiver (or other electronic equipment for that matter) can radiate. But there was no such government rules in the twenties and as more sets came into operation the interference became so bad that radio manufacturers and the public had to voluntarily do something about it. That "something" was called the radio frequency amplifier and is the subject for the next chapter.

Chapter 4

Radio Frequency Amplifiers The Neutrodynes

Edwin Armstrong's regenerative detector worked well for the early broadcast listener until, as mentioned in the last chapter, the neighbor got one. Then the chirps and whistles produced by the neighbor interfered with his reception. Attempts were soon made to reduce this interference by using a radio frequency amplifier to isolate the detector from the antenna. Not only would the amplifier reduce the antenna radiation but the overall performance of the receiver would be increased by the additional amplification.

The basic principles of the triode amplifier are illustrated by the simple circuit, Fig. 4.1a. The "A" battery supplies the proper filament current to the amplifier tube, V1. The rheostat, R2 adjusts the filament current to its proper value. The grid voltage is the sum of the "C" or bias battery voltage and the AC generator. The generator represents the input signal source whether, as in the triode detector, it is an audio frequency voltage, or, in the case of the amplifiers discussed in this chapter, a radio frequency voltage. The "B" battery supplies positive voltage to the plate through the "plate" resistor, R1. The amplifier's output signal is produced by the plate current, Ip,

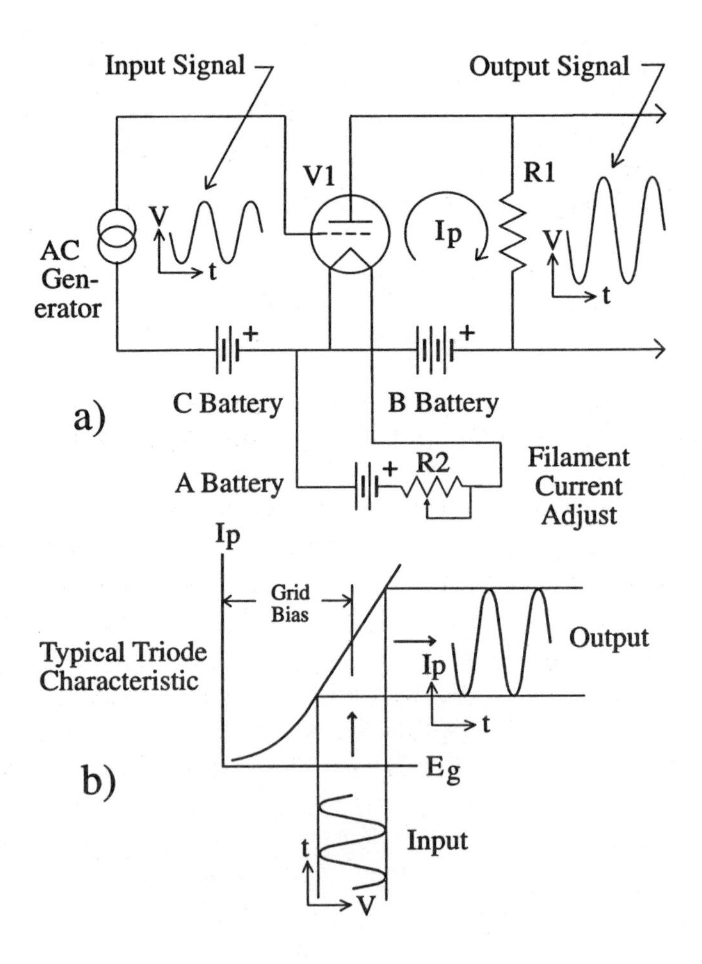

Fig. 4.1 Basic Triode Amplifier

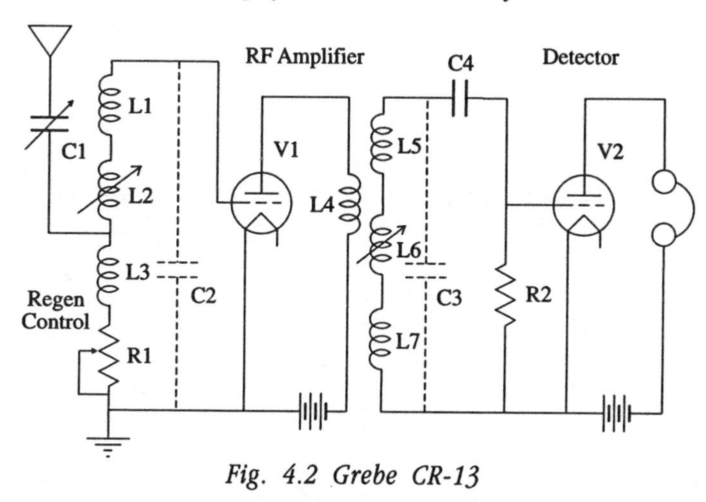

Fig. 4.2 Grebe CR-13

flowing through the plate resistor. The grid control of the electron flow or plate current gives the circuit its amplifying capability.

The control of the plate current by the grid voltage is shown graphically in Fig. 4.1b. The tube's characteristic, the curved line in the graph, shows how the plate current, Ip, varies as the grid voltage, Eg, is changed. The input AC signal, plotted against time, is shown at the bottom of the graph. This signal alternates positively and negatively about the steady grid bias supplied by the "C" battery. The output plate current produced by the input signal is plotted to the right of the graph. For each point on the input signal a corresponding point on the output signal can be obtained by reference to the characteristic curve. The plate current, considerably amplified by the tubes characteristic curve, produces the output signal across the plate resistor, R1.

The basic circuit just discussed was adapted in the early 20's to amplify radio frequency (RF) signals. As illustrated by the 1923 set, the model CR13 manufactured by Grebe, Fig. 4.2, the RF amplifier, V1, used an antenna coupler in place of the basic circuit's AC generator. A second coupler replaced the plate resistor and coupled the amplified voltage to the detector, V2. The antenna tuned by the capacitor, C1, produced the input signal across the tuned circuit consisting of the variometer, L2, and the coils L1 and L3, in parallel with the capacitor, C2. The latter capacitor, shown dashed, was not a separate circuit element but was provided by the inherent capacitance between the turns of the coils.

The signal from the antenna coupler was applied to the grid of the RF amplifier, V1, and produced an amplified voltage across the plate coil, L4, the primary of the second coupler. As in the antenna circuit, the coupler secondary was tuned by the coil capacitance, C3, the variometer, L6, and the coils L5 and L7. The amplified signal was detected by the grid leak detector, V2, which differed slightly from the detector circuit, Fig. 3.3, described in Chapter 3, by having the grid-leak resistor, R2, connected directly to the filament instead of across the grid capacitor, C4. The two circuits are electrically equivalent.

The amount of coupling between the plate coil, L4, and the second tuned circuit, L5, L6 and L7, is one of the critical design parameters for the tuned RF amplifier. Other circuits (see Fig. 4.3 for an example) had connected the plate directly to the secondary coils, L5, L6 and L7. The plate supply battery was then connected to the other end of the coils to supply the necessary plate voltage. But the method used in the Grebe set has since proven itself and became a standard for later tuned RF amplifiers. Yet it seemed novel to the QST editor as late as December 1923 for in an article on the Grebe set he comments that

> the difference between this radio amplifier and the tuned amplifiers we have been accustomed to lies mainly in this plate inductance which is quite loosely coupled to the detector grid circuit and is not tuned at all—in fact, is deliberately made with such a low inductance that it cannot get into tune with any incoming signal. (QST 1923)

The sensitivity of the Grebe receiver was adjusted by the variable resistor, R1, in series with the tuned circuit. Increasing this resistor made the circuit less efficient and thereby reduced the signal to the amplifier. It also had another effect due to unintended coupling between the plate and grid circuits which could result in regeneration. The regeneration increased when R1 was adjusted for high sensitivity and, even though the Grebe set had a non-regenerative detector, it relied on this regeneration for its high performance. As with the regenerative detector the set would start to oscillate at maximum sensitivity. These oscillations could only be prevented when the source of the coupling was identified and its effect neutralized. The development of a stable RF amplifier had to await the solution to this problem.

The instability of the RF amplifier troubled another manufacturer in the design of a new radio, C. D. Tuska, a founder of the American Radio Relay League (ARRL). He developed a circuit designed to overcome this problem. In an article reviewing vacuum tube receiving circuits he describes Armstrong's regenerative detector and mentions attempts to improve its sensitivity by means of

> one or more stages of vacuum tubes coupled together before the detector. These stages are supposed to amplify the radio-frequency before it gets to the detector. If this can be done it is very much worthwhile because the response of the detector increases approximately as the square of the voltage applied to the grid. (Tuska 1923)

When he uses words like "suppose" and "if" he reflects the difficulties of his contemporary engineers in attempting the construction of a practical RF amplifier. He goes on to say that the stages could be coupled by resistance and capacitance but that would not provide a sufficiently high amplification. Alternately a "coil shunted by a variable condenser" could be used to couple the two stages but he complains that this circuit, like the Grebe set, tended to oscillate when tuned. Finally, he describes the plate coil coupling used by Grebe in the CR13. He observes that this circuit can be made to work "without difficulty" at long waves but at shorter wavelengths difficulties are occasioned by the "internal capacity of the tubes themselves".

Here he has hit on the nub of the problem. The plate and grid of the triode tube form a small capacity (about 10 pf for the tubes in use in the 20's) which couples energy from the plate circuit back to the grid circuit. When tuned circuits are used in the plate and the grid, the phase of the plate voltage changes as the set is tuned adding or subtracting to the grid signal. At shorter wavelengths (higher frequencies) the coupling to the grid can be sufficient to cause the circuit to oscillate. The grid-plate capacitance is a characteristic of the tube and can be changed only by tube design and is not under control of the circuit designer.

There had been earlier efforts by tube designers to reduce the plate-grid capacity. Special tubes with very small capacitances were built during World War I by engineers in England and by Latour in France. Tuska describes how the elements were made smaller and the leads were brought out of the tube's glass envelope well away from each other. These efforts invariably reduced the triode tube's performance and the tubes themselves were more costly to produce. There was no comparable effort in the United States which left Tuska and other engineers faced with the problem of stabilizing an RF amplifier using the standard tubes available to them.

Tuska's solution to this problem was embedded in his "Superdyne" circuit. Like the Grebe CR13 the Superdyne, Fig. 4.3, used an RF amplifier followed by a non-regenerative detector but, instead of variometers, the variable capacitors, C1 and C2, were used for tuning. The antenna signal was coupled through L1 to the tuned circuit, L2 and C1, and thence to the grid of the RF amplifier. The plate circuit contained the second tuned

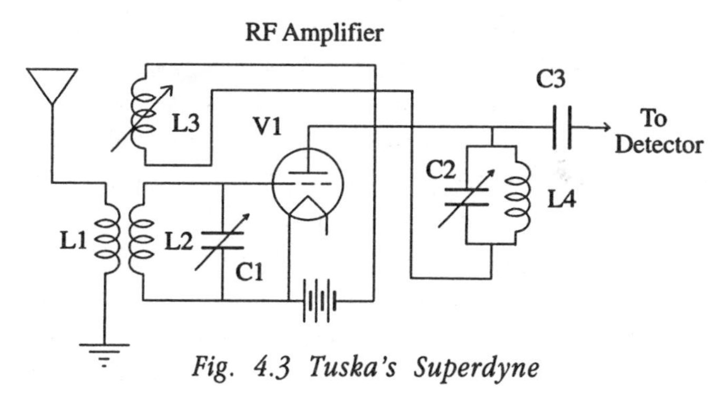

Fig. 4.3 Tuska's Superdyne

circuit, L4 and C2, and the amplified signal was fed to the detector through the grid-leak capacitor, C3. The unusual part of the circuit was the use of the rotating tickler coil, L3, to couple part of the plate signal back to the grid coil, L2. Notice that, if the plate tuned circuit, C2 and L4, were replaced by headphones the Superdyne would bear a striking similarity to Armstrong's regenerative detector, Fig. 3.4.

Tuska first tested his circuit without the tickler coil, L3, and the circuit immediately oscillated due to the grid-plate capacitance. It is interesting that he and his associates were frustrated as he writes:

> It was evident that we *must* use resonant circuits and it was further apparent that the minute we did use resonant circuits the tubes would start to oscillate and spoil everything. We were in a "vicious circle".

This, of course was the difficulty that plagued all triode RF amplifier designers. Then Tuska came up with his idea:

> All that is necessary is to put in the conventional Armstrong feedback but *feed the energy back in the reverse direction or negatively*.

As Armstrong had used a tickler to provide positive feedback in his regenerative detector so did Tuska use a tickler coil, L3, arranged as in a regenerative detector but made to couple with the opposite polarity. The signal from the tickler coil would then nullify the signal from the grid-plate capacity and the circuit would be stable. When the tickler was not adjusted to completely prevent oscillations it provided regeneration and consequently a high degree of amplification as Tuska later found out:

> By carefully adjusting the reverse feedback against the positive capacity-feedback one can get astounding degrees of amplification.

Unfortunately for Tuska his circuit had a serious drawback. As is pointed out in Morecroft's textbook on radio engineering, oscillations in an RF amplifier can be prevented if

> Another, and opposing, voltage can be introduced into the grid circuit by an electromagnetic coupling between the grid and plate circuits. (Morecroft 1927)

Which is precisely what the Superdyne accomplished. But here's the kicker, again quoting from Morecroft:

> Such an expedient can be expected to work over a comparatively narrow frequency band, however, as it is not possible to just balance a capacitive feed-back by magnetic feed-back through a wide range of frequencies. The magnetic feed-back must be made adjustable if such a scheme is to be most effective, and the operator will have to change the magnetic coupling as he changes the tuning condenser.

Perhaps with one RF stage and two tuning controls it was not too difficult to tune Tuska's set and, at the same time, adjust the tickler coil. But, later, when RF amplifiers with more than one RF stage were constructed it would have been very difficult for the operator to keep all the ticklers at the proper setting.

Before Tuska made the Superdyne the proper solution of the oscillation problem was invented by Hazeltine, Professor of Electrical Engineering at the Stevens Institute of Technology. His research covered many aspects of radio and radio receivers. He must have been a very practical professor as he is shown hard at work on one of his receivers in an illustration in Radio News, May 1923. His idea for nullifying or "neutralizing" the tube's grid-plate capacity originated in 1919 when he was a consultant for the Navy on the design of a receiver for long wavelengths. Unwanted short wave signals would pass through the small capacitance between the primary and secondary of the normal antenna coupler. So one of the design requirements for the new receiver was to eliminate this capacitance coupling. But even with metal shielding there still existed the unwanted capacitance between the primary and secondary coils. Hazeltine neutralized the effect of this capacitance by means of an auxiliary capacitor. This capacitor was placed in the circuit so as to couple an equal an opposite signal to the antenna. This voltage cancelled the effect of the stray capacitance and satisfied the Navy's design requirement.

It didn't take long for Hazeltine to apply this principle to the design of an RF amplifier. His circuit, the "Neutrodyne" was described in Radio News May 1923 and is shown in Fig 4.4a. The interstage coils are wound so that the secondary voltage opposes the primary voltage. The secondary voltage is then coupled back to the grid by the "neutralizing" capacitor, C2, and opposes the voltage coupled back to the grid by the tube's capacitance. This small "neutralizing" capacitor is adjusted so that the signal passed by the grid-plate capacitance is nulled out and its effect on the circuit is neutralized. The amplifier then operates as though there were no grid-plate capacitance and it has no tendency to oscillate. Because a capacitor is used to balance out another capacitor the neutralization is independent of frequency. Thus Hazeltine's circuit had the great advantage that the neutralizing capacitor could be adjusted once and for all at the factory and did not have to be changed as the set was tuned, which was not the case for Tuska's circuit.

In order to achieve proper neutralization, the neutralizing capacitor, C2, Fig. 4.4a, must be adjusted to a value approxi-

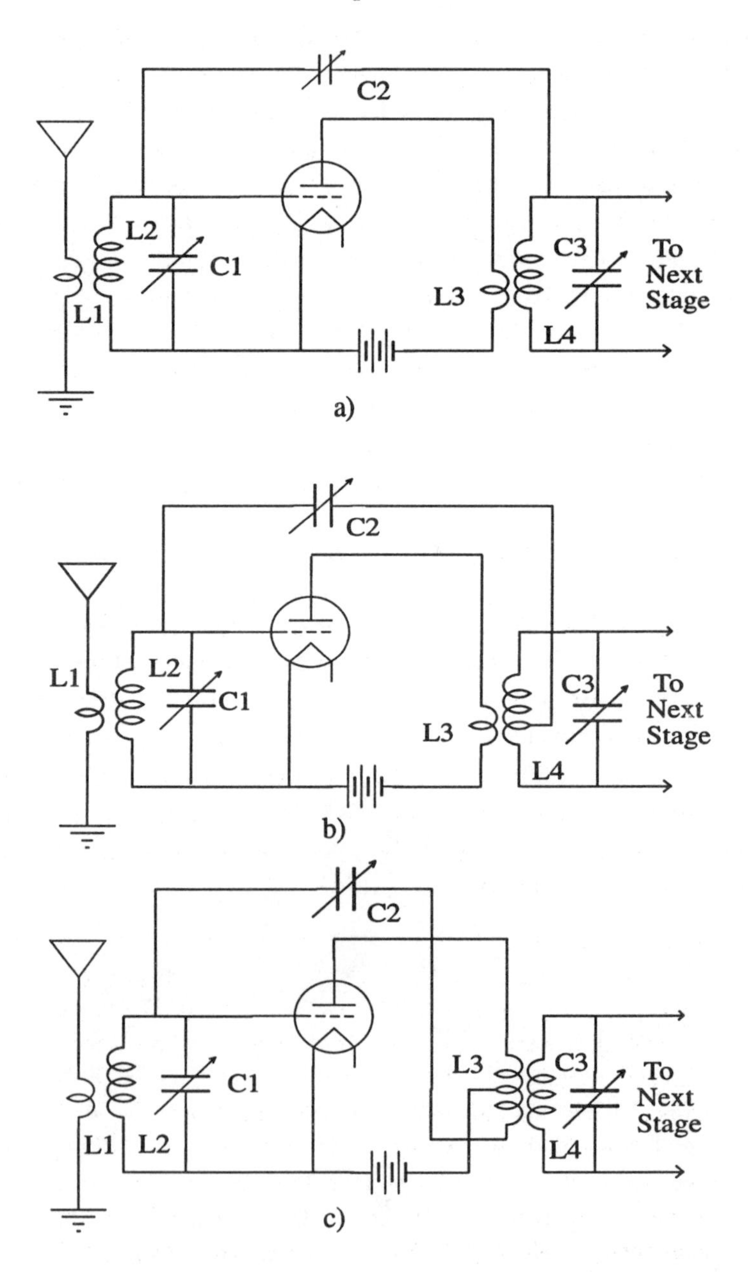

Fig. 4.4 Neutrodynes

mately equal to the tube's small grid-plate capacitance (about 10 pf). Such small capacitances were made by placing two pieces of insulated wire inside a brass tube (Plate 13). The tube was slid back and forth to adjust the capacitance for proper neutralization. These capacitors were difficult to adjust so that when the Neutrodyne circuit came to be put to use in commercial receivers the neutralizing capacitor was moved to a tap on the secondary coil, Fig. 4.4b. This change allowed the use of a much larger neutralizing capacitor with rotating plates which made the adjustment much easier. By 1926 the Neutrodyne circuit had been changed again to its final and most popular configuration, Fig 4.4c. A separate coil, L3, (sometimes wound continuously with the plate coil) supplied the opposing neutralizing voltage.

Once amplifiers were stabilized by the Neutrodyne circuit engineers could attend to the optimum design of the interstage coils which, taken together, are called an "RF transformer". As the article on the Grebe CR-13 mentioned, the coupling between the plate coil and the following tuned circuit was made "loose". But how loose? This question and other practical problems of the design of the RF transformer was investigated thoroughly by two engineers at the Cruft Laboratory at Harvard University. One of them, Glenn L. Browning, describes the development as follows:

> In August of 1923, Mr. F. H. Drake showed the writer some mathematics on a vacuum tube used as an amplifier in conjunction with a tuned radio frequency transformer, stating that there was a lack of efficiency in most R. F. transformers, and suggesting a thorough mathematical analysis of the circuit which, together with laboratory measurements, might throw some light on the right constants which were necessary for maximum amplification. Accordingly he and the writer worked together. . . for almost a year with the result that considerable information was collected. (Browning 1925)

Their analysis was able to predict the circuit values that would give the maximum amplification. They then tested RF transformers constructed in the "ordinary manner", that is, with the primary and secondary wound as single layer coils with the one inside the other. They found that actual laboratory performance was far below that predicted by their analysis. After some investigation they discovered that the source of the problem was the inherent capacity between the primary and secondary windings. This capacity coupled signal voltage to the secondary that opposed the magnetic coupling from the primary.

They came up with a new RF transformer design that minimized this capacity. The primary or plate winding was wound on a "wooden disc with a groove cut in it . . . so as to fit snugly inside the tubing on which [the secondary tuned coil] is wound." This "slot coil" was placed at the ground end of the secondary so that most of the stray capacitance is shunted to ground. The resulting design was called the "Regenaformer" and the set they had built in their laboratory became known as the Browning-Drake set.

They licensed National to manufacture their transformers and National sold kits to experimenters. The National coils were combined with the new "straight-line" frequency tuning capacitors the company had recently developed (Plate 5). These capacitors made tuning easier by using specially shaped plates so that the tuning dial readings were proportional to frequency. Thus stations at the high frequency end of the dial were not crowded together.

The schematic of Browning-Drake's set, Fig. 4.5, is still today an example of a straightforward and elegant circuit. The RF stage used Neutrodyne neutralization with feedback from a tap on the tuned coil, L3, through the neutralizing capacitor, C3. The detector's efficient RF transformer used a slot coil, L2, for the plate winding and located it at the ground end of the

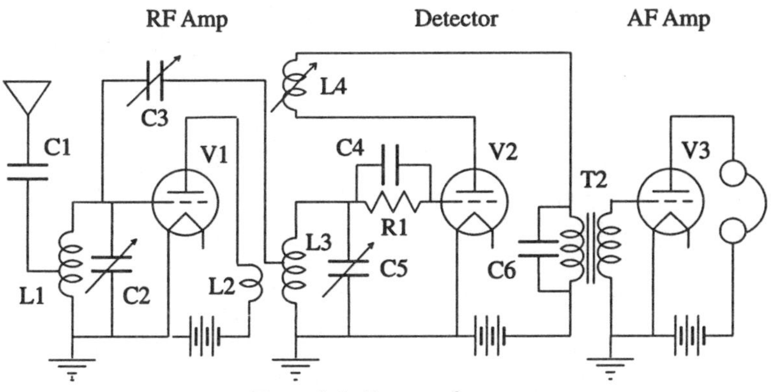

Fig. 4.5 Regenaformer

secondary, L3. Regeneration was supplied by the tickler coil, L4, as in Armstrong's detector, Fig. 3.4.

The receiver thus combined the work from four of the early radio geniuses: Armstrong, Hazeltine, Browning and Drake. The result was a highly sensitive receiver for its time. Reports from users said that it outperformed sets using multiple RF amplifiers and non-regenerative detectors. But a regenerative detector is always tricky to adjust and is difficult for the general public to use. Later sets with many stages of RF amplification did away with the need for the regenerative detector. But, even without Armstrong's detector, the Browning and Drake's RF transformer was applicable to all RF amplifiers and their design principles are still being followed at the present time.

Chapter 5

The TRF's Multi-Stage RF Amplifiers

The Browning-Drake circuit provided good sensitivity but it wasn't really acceptable as a radio for general public use. The adjustment of the regeneration control required some practice because the setting changed as the set was tuned. This made it difficult to tune in long distance stations. The regenerative detector was therefore discontinued by most manufacturers and the resulting drop in sensitivity was made up by using additional stages of RF amplification.

The idea of cascading RF amplifiers was not new in the 20's. Almost a decade before, in 1913, E. F. W. Alexanderson, the same engineer who had developed the high frequency alternator at GE, patented a multistage amplifier coupled by tuned circuits (Archer 1938 p.120). Radios using tuned radio frequency (TRF) coupling became so popular during the twenties that the sets themselves were called TRF's for short.

A simplified schematic of a typical two stage TRF amplifier, Fig 5.1, shows how Alexanderson's scheme was used with Hazeltine's Neutrodyne circuit. Two Neutrodyne stages, V1 and V2, and a grid leak detector, V3, were coupled in cascade by identical RF transformers tuned by the capacitors, C3, C4, and

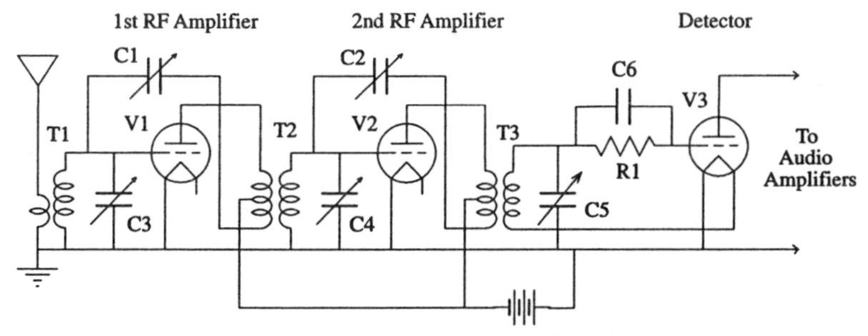

Fig. 5.1 Two Stage TRF

C5. The two amplifiers were neutralized by capacitors, C1 and C2. Each of the three tuning capacitors required a control knob on the front panel so that they could be tuned to the same frequency. When searching for a station the operator had to turn the knobs together a little at a time. This was difficult enough for three knobs but a three stage TRF set required four knobs which were even more difficult, if not impossible, to tune together. For this reason, commercial TRF radios usually used only one or two RF stages.

Nowadays with the modern high gain RF circuits shielding is put in as a matter of course. But it wasn't so for early TRF's. They were all built on wooden or bakelite chassis in the tradition of the experimenter's "breadboard". A typical layout of the component parts, Fig. 5.2a, followed the schematic diagram from left to right beginning with the antenna RF transformer, L1, mounted directly behind the first tuning capacitor, C1. Then following directly to the right was the first RF amplifier tube, V1, and its RF transformer, L2, and capacitor, C2. Next came the second RF stage, V2, and the third transformer, L3, and capacitor, C3. The detector, V3, was followed by an audio transformer, T4, and the audio frequency amplifier, V4. When more than one stage of audio frequency amplification was used these stages followed further toward the back and then, in some cases, reversed the order along the back of the set (Plate 20). The rheostat, R1, controlled the filament current of the RF amplifiers

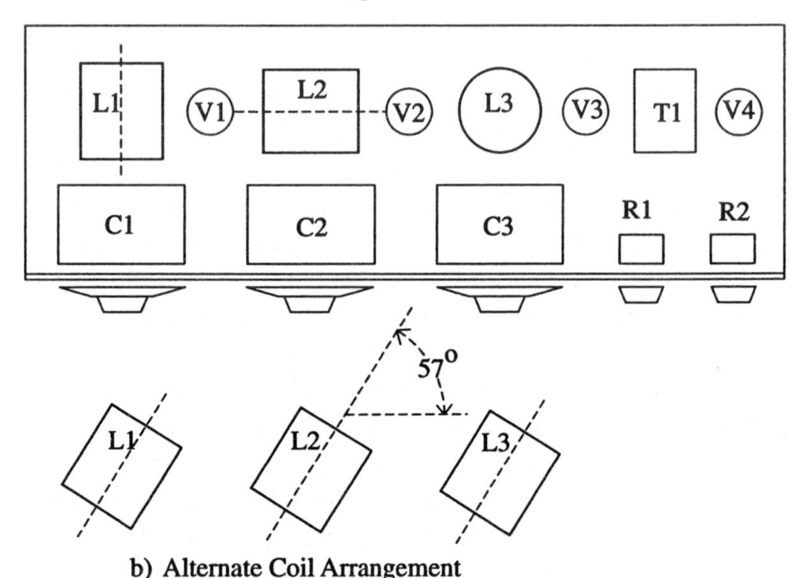

Fig. 5.2 TRF Layout

and was used as a volume control while the rheostat, R2, set the filament current for the other tubes. The large tuning knobs were simply calibrated with divisions from 0 to 100.

The RF transformers, usually wound on "hard rubber" cylinders, had to be carefully placed to avoid undesirable capacitive and magnetic couplings between them. Coupling was at a minimum when the axes of the transformers were placed at right angles to each other. The three transformers of the two stage RF amplifier can be placed with their axes at right angles to each other as shown by the dashed lines, Fig. 5.2a. Each axis is parallel to one of the three space dimensions, two horizontally and one vertically (Plates 20 and 22). However, since the fourth dimension is not available, it's not possible to place the four RF transformers of a three stage amplifier at right angles to each other. But Hazeltine came up with an alternative arrangement which could be used with any number of stages.

He developed a mathematical theory of electromagnetic coupling between RF transformers. He found that the coupling was minimum when they were placed with their axes aligned at a certain angle to a common center line. He calculated the proper angle to be the one whose tangent is the square root of 2, that is 54.7 degrees. This alternative configuration, Fig. 5.2b, shows this alternate alignment for the two stage TRF. Morecroft (1927) repeats this information in his book "Principles of Radio Communication" but does not give the mathematical proof. This technique proved itself in practice and was used in the construction of many TRF's (Plate 13).

Another design problem in early TRF's was the incorporation of a "volume" control. Usually the audio amplifiers after the detector were allowed to run "wide open" without a potentiometer gain or volume control as we now have in our stereos. A common method of gain control was by adjusting the RF amplifier's filament voltage by means of a filament rheostat. Alternately a variable resistor was inserted in the antenna tuned circuit to decrease its efficiency. A scheme applicable to a set which uses a tuned loop antenna instead of an antenna transformer was used by Music Master. The volume control in their shielded four stage Neutrodyne receiver consisted of an auxiliary tuning capacitor placed across the loop's main tuning capacitor. When this capacitor was increased the loop was detuned sufficiently to reduce the set's overall sensitivity and reduce the volume.

All these volume control methods provided, for the same listening volume, a constant voltage to the detector and the following AF amplifiers. Thus very large signals were prevented from overloading these amplifiers. An interesting description of how overloading occurs is given by Laing in his article on the Music Master set. He says that

> This feature is one that has not been emphasized sufficiently until recently. Anyone who is familiar with the common

> "characteristic curve" of a vacuum tube should know that, when the fluctuation of grid voltage becomes greater than the amount of voltage represented by the straight portion of the graph, distortion will result. . . The result is a peculiar form of distortion in which almost all sounds have the same intensity; and in which the most powerful impulses are given a sound best described as "mushy". (Laing 1926)

"Mushy" may not be the best expression when one hears the raucous sound from a teenager's transistor with the volume turned all the way up but the cause of the problem is the same.

In 1922 F. A. D. Andrea (FADA) was marketing a kit based on Hazeltines circuit and in 1923 was offering a complete receiver, the "One-Sixty". Using three stages of neutralized RF amplification and a NON-regenerative detector it was a marked improvement of the squealing and howling of the earlier regenerative sets. In Radio Broadcast magazine the chief engineer of FADA, Kimball Stark, describes the history leading up to this circuit and the set itself with the concluding comment:

> The selectivity of the receiver is great and yet because no regeneration occurs it is possible for even the novice broadcast listener to adjust the three dials quickly and receive concerts with great clarity. Dial settings for various stations read like football signals, and to be able to have the women folk turn the dials to prearranged settings, throw in the filament switch and pull in broadcasting stations 1500 miles away, is a feat that even some of the older radio "nighthawks" envy. (Stark 1923)

So came into being one of the first of many TRF's with three tuning knobs with dials calibrated simply 0 to 100. Settings like 43-38-50 for KDKA were "football signals" indeed but they appealed to the "night hawk" who listened to distance stations at night when the reception was good. The author remembers the thrill of long distance reception when he was a young man. On vacation up in the mountains and away from city noises, the car radio would bring in far away stations and he still tunes his stereo to 1070 AM at night-time to receive KNX, Los Angeles, all the way up here in Oregon. This feat that the "women folks" accomplished with "football signals" is now done from the luxury of the arm chair with digital tuning and a remote control.

Yet the TRF was still new in 1924 and all the problems were not yet worked out. Healdon Starkey, a Zenith engineer, gives his opinion of the state of the art in 1924 saying that

> For the past two years radio-frequency amplification at the broadcast and amateur waves has been approved and rejected, knocked and boosted, canned and re-instated until the average person has come to the conclusion that "radio-frequency amplifiers are all right when they work, but they only work when they feel like it", whereas the real trouble all along has been that the radio-frequency amplifier was only partly developed. (Starkey 1924)

He then goes on to make light of the use of "tuned transformers" having high losses, that "marvelous destroyer of religion" in which both plate and grid circuits were tuned, and finally the "reversed tickler" (Tuska's Superdyne) and the "Neutrodyne types".

But the TRF's got better and better and an example of a good design was produced by that early manufacturer of regenerative sets, the Grebe Company, when they came out with a TRF of their own at the same time Starkey made his disparaging comments. A special editorial article in the October, 1924, issue of QST announced their "Synchrophase" MU1 receiver.

The early Grebe sets had mounted some of their controls with the shaft axis vertical behind the panel with the edges of the knobs protruding through slots. This allowed their adjustment by means of the operator's thumb which, if not too convenient for some, was indeed novel. The Synchrophase receiver carried this arrangement over to the tuning dials themselves by mounting the tuning capacitors with their shafts vertical. The set's two RF stages required three tuning capacitors with three thumb control tuning dials. Fine tuning was accomplished with three mechanically connected fine or vernier controls below the main dials (Plate 15). It seems that a good "Variable condenser" was hard to come by in the early days and as the QST editor said it was

> one of QST's pet topics and in fact our agitation has been the main cause for the present epidemic of "low loss" condensers. (Incidentally—we invented that term "low loss").
>
> (QST 1924b)

He goes on to praise the Grebe units "straight line frequency" tuning capacitors. Early variable capacitors were made with their moving plates shaped in a semi-circle so that their capacity varied linearly with their rotation (Plate 8). Since the frequency of a tuned circuit varies as the inverse square of the capacitance this resulted in radio stations being bunched together at the high frequency end of the dial. The straight line frequency capacitors solved this problem by having specially shaped plates so as to space the frequencies more evenly across the dial (Plates 5 and 9). The popularity of these capacitors caused them to be standard for tuning applications and the "straight line" shape is still used in today's receivers (except the digitally tuned units of course).

The construction of the RF transformer coils is another unique feature of the Grebe Syncrophase. As the QST editor puts it

> Builders of neutrodynes are familiar with the care necessary to prevent magnetic feedback between the coils in different stages of the amplifier. In the "Synchrophase" this difficulty is avoided in a different manner by making the coils of a special shape which has practically no magnetic field outside of the coil itself. (QST 1924b)

Grebe called these special RF transformers "Binocular Coils" and they were developed by R. R. Batcher, Research Engineer with the company. In order to get the maximum amplification per stage and prevent oscillations he investigated methods to prevent inductive coupling. He had considered a toroidal or "doughnut" coil but he "discarded" it without giving a reason. At about the same time Thorola was making sets with toroidal coils so Batcher may well have wanted something different (Plate 6). Today's ferrite core toroids confine their magnetic field well but Batcher would have had to wait about 30 years for that development. So he took two solenoid coils and placed them side by side, wound in the same direction and connected in series so that their mutual coupling aids each other (Plate 17). The magnetic field of one is therefore in the opposite direction of the other. A magnetic field from a nearby coil or an electromagnetic field from a nearby transmitter would therefore pass through both of the coils in the same direction, the induced voltage in one cancelling that in the other. The binocular coils were also made highly efficient by placing the primary or plate coils inside near the ground end of the tuned secondary. The lessons of Browning Drake had sunk in.

There were really no new circuits used in the Grebe Synchrophase. As Batcher points out, "the interesting thing about the Synchrophase is not the circuit but the design of the parts used in it." By further careful placing of the components he obtained a set without oscillations, high gain and good selectivity—the ultimate aim of receiver design.

Although the basic TRF circuit, Fig. 5.1, provided good performance in radios such as the FADA and Grebe sets it has one serious weakness. The RF transformer efficiency changed as the set was tuned to different frequencies. As the tuning capacitor was increased to tune to lower frequencies the efficiency of the transformer decreased causing a loss of sensitivity at the low frequency end of the band. Many designers corrected for this change in sensitivity by placing a resistor in series with the grid of each RF stage. The resistor lowered the amount of high frequency amplification to match that at the low frequencies resulting in less of a change from one end of the dial to the other but at the cost of a reduced overall amplification.

An ingenious circuit designed to solve this problem was designed by K. E. Hassel of the Chicago Radio Laboratory and Zenith Radio Corp. He attacked the problem of overcoming

> the marked decrease in the transfer of energy from one stage to the next with the increase of wavelength . . [and he] calmly did the thing what is so obviously the simplest and most efficient way of meeting the difficulty that everyone who sees it . . . says "Why didn't I think of that!" . . .He simply took a portion of the plate coil and mounted it on the condenser shaft so that it rotates inside the grid coil of the succeeding stage. (Starkey 1924)

As the tuning capacitor increased to tune to lower frequencies the rotating coil, operating like a variocoupler, turned to increase the coupling of the RF transformer thereby equalizing the gain.

Hassel's circuit was incorporated into the "Super-Zenith" and it's performance seemed to be very good as

> On at least one occasion during the summer it put KDKA on the loudspeaker with dancing volume without any kind of aerial or ground. (Starkey 1924)

A much more sophisticated solution to the problem was devised in the same year by Carl E. Trube a consulting engineer for Thermiodyne. He devised a circuit that, as the frequency changed, varied the coupling between the RF transformer's primary and secondary without the use of moving coils. The amplifier stages were not neutralized and the goal of the design was to make the amplification constant at a value below the threshold of oscillation. Thus, lacking neutralization, the ampli-

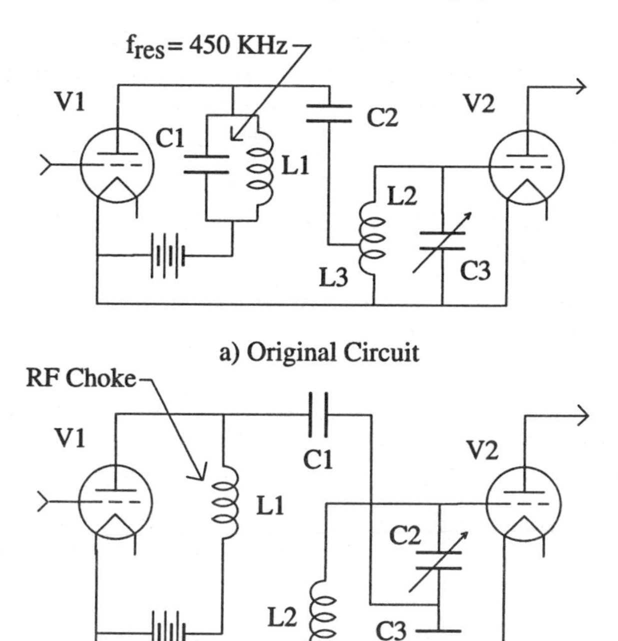

Fig. 5.3 Thermiodyne Circuit

fication was kept at the highest value possible throughout the broadcast band.

Thermiodyne used this circuit in their TF-6 receiver, Fig. 5.3a, (Wheeler and MacDonald 1931). The normal coupling or primary coil of the RF transformer is not used. Instead the coupling is made through capacitor, C2, to a tap on the secondary tuned circuit. The additional coil, L1, is wound on a separate form not coupled to the secondary, L2 and L3. The values of the capacitor, C1, and the inductance L1 are chosen so that they resonate at a frequency of 400 to 450 KHz, slightly lower than the broadcast band. When the secondary is tuned by C3 to the lower broadcast frequencies this resonance increases the effective coupling through L3 and increases the gain of the circuit.

A later circuit developed by Trube, Fig. 5.3b, is different in that the coil, L1, is now wound to have a high inductance and just serves to supply plate voltage to the tube. The coupling capacitor, C1, is not directly coupled to the tuned secondary, L2 and L3, as in the earlier circuit. Instead another capacitor, C3, is inserted at the ground end of the tuning capacitor, C2, to connect to L3. The operation of the circuit is more complicated than the previous one but can be explained by the fact that, at lower frequencies, the tuning capacitor has more capacitance and more current flows through it to the secondary, L2, thereby increasing the effective coupling and the gain of the stage.

These circuits were arrived at experimentally but later Harold Wheeler and W. A. MacDonald of the Hazeltine Corporation made an extensive theoretical analyses of many types of coupling circuits (Wheeler and MacDonald 1931). Their work gave a firm theoretical foundation for the design of RF transformers used in radios of the 30's.

Chapter 6

The "Lossers"

The five years, 1923 to 1927, saw a proliferation of receivers with tuned radio frequency amplifiers that did not use the Neutrodyne circuit. The incentive for this development was to avoid payment for Hazeltine's patent license. In so doing the engineers, in many cases, developed a very stable receiver with few internal adjustments.

Possibly the most famous of these sets was made by Arthur Atwater Kent, founder of one of the largest and most successful manufacturers of receiving sets that bears his name. Atwater got his start in automobile ignition components where he not only made a number of inventions but also, in the process of manufacturing operations, became expert in making bakelite moldings. These moldings required expensive steel molds made in his factory and he turned this expertise into the manufacture of radio parts in 1922. He produced the famous Atwater Kent "breadboards" having handsome bakelite parts mounted on a wooden board in the style of the home-built sets.

The word, breadboard, has remained in our language and now means a "quick and dirty" working model of a circuit or even software. However, there was nothing "quick and dirty" about Atwaters breadboards. In fact they were looked upon as "work of art". The public accepted his breadboards even though they were without front panels or cabinets. The bakelite molded variometers, tuning capacitors, and tube assemblies contained all the component parts and were easily placed on a wooden base and hooked up with the minimum of wiring and were sold as either kits or fully assembled.

Later the public demanded sets for the living room so Atwater made complete TRF receivers in cabinets using many of the same bakelite parts in his breadboards Plates (22 and 23). He was able to make these sets attractive, with crackled paint metal panels and walnut cabinets, and he had sold a million sets by December 1925.

Atwater did all this without a license from Hazeltine. Why didn't his sets oscillate? In order to answer this question a review must be made of the other possible ways that may be used to reduce the effect of the tube's grid-plate capacity.

First, the neutralization can be accomplished by carefully arranging the tuned circuits so that stray capacities will provide the necessary neutralizing capacity. Thus Hazeltine's circuit was used but hidden from view. This method worked even better if the amplification or gain of the amplifier stage was reduced.

The gain could be reduced by either using inefficient coils, by placing resistors across the tuned circuits or coupling the coils to a resistive circuit. All these methods reduce the efficiency of the circuit by introducing elements that provide a "loss" in performance. In many early designs, the amplifier gain was reduced from the its ideal value by not following the RF transformer design principles laid down by Browning and Drake. The editors of the amateur magazine, QST, dubbed them the "Lossers".

The resulting overall amplification remained sufficiently high for ordinary radio listeners although they were not as sensitive as a well-designed Neutrodyne. They also were easily aligned as there were no neutralizing capacitors to adjust. This saving in labor and the costs of the capacitors themselves, not to mention the license fees which could amount to as much as 7.5% of gross sales, made the sets inexpensive to manufacture and the companies could undercut the competitor's price. In this way, Atwater and many other radio engineers used the simple expedient of not optimizing the radio frequency amplifier design parameters. Yet they succeeded in producing a reliable, straight forward set that gave trouble free performance to a great many customers.

In his article in Radio Broadcast, November 1927, J. O. Mesa, engineer for the Freshman Co., reviews the state of the art of receiver design. He describes the "losser" circuit as follows:

> A method that has been used to maintain the circuits in a two- or three- stage radio-frequency amplifier free from oscillations is that of including resistances in grid circuits of the amplifying tubes. (Mesa 1927)

He correctly points out that this method of "holding down" an amplifier is equivalent to placing a resistor in series with the tuned coil. This reduces the selectivity and efficiency of the tuned circuit, reducing the amplification of the RF stage. In modern terms the "Q" of the coil has been reduced. Mesa then points out that:

> Under the usual conditions obtaining in a high-amplification circuit, the value of the stabilizing resistance is somewhat critical. If the correct value is exceeded, the overall voltage-gain of the radio-frequency amplifier is considerably reduced, while if the resistance is too small, the circuit oscillates violently.

He mentions that his experience in the mass production of sets at the Freshman Co. showed that, even with the use of close tolerance resistors, that a

> . . .difficulty occurred from oscillation or poor selectivity and lack of amplification too frequently for comfort.

So the designers more than likely erred on the side of poor selectivity and low amplification.

The earlier Freshman sets, built before Mesa wrote his article, did not use a series grid resistor in the radio frequency stages. In 1923 they produced their "Masterpiece" which was designed to undercut their competitors by the use of the lowest cost circuits and construction. The stability problem was neatly solved by mounting the tuning coils directly in back of the capacitors. Not only did the capacitor become the mounting plate for the coil but its frame induced losses in the coil. The frame, built of steel, a conductor of electricity, has eddy currents induced in it and acts as a continuous shorted turn coupled to the coil. Because the steel is a poor conductor, the effect is the same as placing a resistor in series with the coil. The same effect as the grid resistor but without the expense of the resistor!

Freshman provided an ingenious dial with an internal gear stamped out of brass and mounted it on the front of the tuning capacitor. The whole assembly comprising the radio frequency transformer (coil), capacitor, and dial was easily and quickly installed in the receiver (Plate 11). Three of these units, for the two radio frequency amplifiers, plus the tube sockets, audio transformers and rheostats make up the entire receiver. Easily assembled on a brass panel with mock wood grain finish its no wonder it was cheap to build!

A not so well known manufacturer, Appleby, used a similar technique by either mounting the coils near the tuning capacitors stators or close to a steel plate sub-panel. In an article in Radio News Rowe (1926b) says that

> These losses are produced by the proper placing of the second radio-frequency transformer in relation to the metal shielding. It is not possible to arrive at this position mathematically but only through experimentation.

It is also pointed out as a feature of the circuit that the primaries of the radio frequency transformers are wound in the middle of the secondaries. This in the face of Browning-Drake's work who showed that they should be wound at the bottom to minimize losses.

The introduction of loss took many forms. The Pfanstiel set in 1926 uses two stages of radio frequency amplification and

> . . .the losses intentionally introduced into the circuit by the use of "doped" spider-web coils, combine to suppress any tendency to oscillate; and broaden the tuning enough so that the three condensers can be used with a single control, without the necessity of auxiliary verniers. (Griffin 1926)

The "spider web" coils were wound in a flat spiral like a spider web. Presumedly the "doping" is some sort of high loss coating or cement.

Besides the manufacturers already mentioned, Atwater Kent, Freshman, and Pfanstiel, many other manufacturers used grid resistors in their RF amplifiers. In January, 1927, Radio News says of the Federal-Brandes set:

> The radio-frequency stages are of the resistance-stabilized type, and do not betray any signs of oscillation anywhere over the tuning scale. (Radio News 1927)

So it seems that the technique had by that time become well known and had no need of explanation.

Chapter 7

The Mechanics of One-Knob Control

In the early 20's the TRF's were built on a "breadboard" with their electrical components placed left to right, in order of their function, Fig. 5.2. This "traditional" configuration allowed the amateur builder to follow the schematic diagram left to right as he wired his set. It also gave the necessary mechanical separation between stages to reduce the unwanted electrical coupling between the components to a tolerable level. Later, when the number of RF stages and controls increased a panel was added to the breadboard. When the sets became parlor fixtures the complete breadboard with panel was housed in an attractive cabinet.

The tuning controls necessarily followed the pattern of the component layout (Plate 12). A simple dial knob, or one with a mechanical vernier, was mounted on each of the capacitor shafts across the wide front panel. As previously discussed a two-stage RF amplifier required three controls. All three had to be tuned to the proper position to select a station. The numbers on the dials did little to help when searching and each dial had to be slowly turned a little at a time trying to keep them in step. For a time it seemed that adjusting these three dials was a skill that went along with owning a radio, like learning to drive a car, and was shown off by the proud owner. Once the favorite stations were found, the dial readings could be written down "like football signals" so that everyone, even the "women folk" could readily tune to these settings.

By 1924 the average radio listener wanted simpler tuning and the ideas for controlling all the tuning capacitors with one knob came into being. Of course the old-timers resisted this innovation at first, partly because some performance was sacrificed and partly through tradition—"Why were we born with two hands if not to use them". But a set with three TRF stages required four knobs and we sure don't have four hands.

These "one knob" sets differed both mechanically and electrically. Mechanically, different methods were used to couple all the tuning capacitors together. Electrically, different circuits were used to make the tuned circuits track each other as the capacitors turned in synchronism. This chapter will describe the ingenious mechanical mechanisms invented for one-knob control leaving the electrical problems for the next chapter.

In the few years between 1924 and 1927 there was a sudden rise in the production of new mechanical ideas for one-knob tuning. The magnitude of this sudden rush of activity can be inferred from the patent office statistics. Over this period the number of patents for single control mechanisms rose dramatically from less than 10 to over 100 (Harrison 79). The designers were not only spurred on by the competition but also with the great popularity of their sets with customers.

This demand may have been given a further stimulation by the introduction of the superheterodyne by RCA in 1924 (Harrison (1983). The "superhets" had only two knobs, one for tuning and one to trim up the antenna. But the inherent operation of the circuits did not allow these controls to track well. It would seem that on the basis of simplicity of operation a well designed one knob TRF would have been much easier to use. It wasn't until the late 20's that methods were invented to make the tuning of the superhet possible with one knob.

Long before there was an "average" radio listener a very early attempt in "one-knob" control was developed during World War I. The receivers at that time, like the crystal sets described in Chapter 2, were tuned by using tapped coils. Taps placed close together were connected to a rotary switch for fine tuning while taps placed farther apart were connected to a coarse tuning switch. Then, as today, the Navy wanted to scan the radio frequencies for enemy transmissions. An operator attempting to scan through the entire range of the receiver would necessarily have to alternately use the coarse and fine tuning switches. With each setting of the coarse tuning switch the fine tuning would need to be turned through each of its positions until all the combinations of switch positions had been selected. A continuous search then required that the entire process be repeated all over again.

A one-knob solution to this problem, which seems to be closely related to ideas incorporated in mechanical counters, was invented by Roy E. Thompson and described in his paper in the Proceedings of the Institute of Radio Engineers (Thompson 1919). His idea was to gear the two rotary switches together so that the coarse switch would make one step for each revolution of the fine switch. Then the operator could tune through the entire band by operating the fine switch alone. His receiver, the "Unitrol", was perhaps the first one-knob receiver. He went on to drive the switches with an electric motor to produce an automatic scanning receiver, an early forerunner of the scanners that are now available to everyone for listening to police and aircraft communications.

By the time one-knob control receivers were being developed tuning was no longer being done with tapped coils but with either variometers or variable capacitors. Although the designers of the TRF receivers traditionally mounted these units from left to right the first one-knob sets broke from this tradition and arranged the capacitors front to back on a single shaft. This configuration was universally used all through the 1930's up to the present day. It is difficult to know which inventor should be given the credit for being first with this arrangement. As far as using a single shaft front to back the Magnavox TR-5 was the first. This set was designed by the two engineers responsible for the horn loudspeakers which made Magnavox famous. As told by Douglas (1989), Peter L. Jensen was a Danish engineer brought to California to work with the Federal Telegraph Company. He and his fellow engineer Edwin S. Pridham formed the Magnavox Company in 1917. Later they expanded the company's products to include radios and in 1924 foresaw the need for one-knob tuning. In place of the usual tuning capacitors they used specially designed "flat" variometers. The three variometers required by the two stage TRF circuit were mounted on a single shaft which was driven directly by a dial on the front panel. Even today, the construction looks very modern. The tubes followed the practice of the 30's by being placed front to back alongside their corresponding variometers.

The first single shaft arrangement driving three tuning capacitors on one shaft was invented by Paul A. Chamberlain and Douglas De Mare of the Mohawk Electric Corporation (Harrison 1979). This arrangement was the true forerunner of today's multiple tuning capacitors. Chamberlain received a patent for his idea on July 2, 1924 and it was incorporated in the Mohawk A5 receiver. The 3-gang tuning unit incorporated adjustable metal strips on the side that served as trimmer capacitors to help equalize the three tuning capacitors. These strips were the forerunner of the mica compression trimmer capacitors now used in later units (Plate 10). The layout of the tubes and coils still followed the traditional left to right arrangement which resulted in long wires to the capacitors which not only looked untidy but must have had adverse effects on the

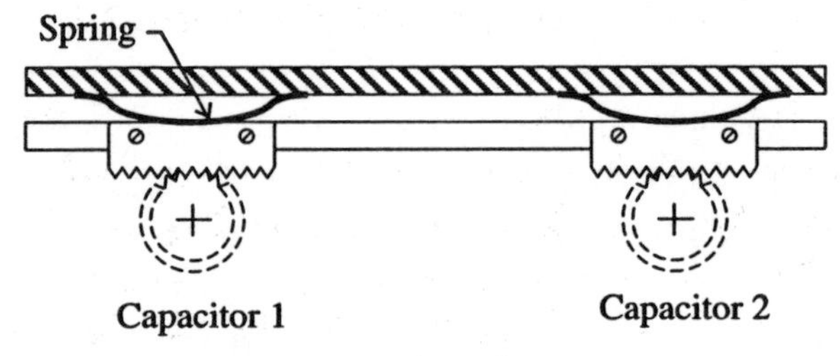

Fig. 7.1 Rack and Pinion

electrical performance. The Magnavox unit was much more modern in this respect. Later, in 1926 Mohawk re-arranged the tube layout front to back in keeping with modern practice.

Many designers used a single shaft but ran it from left to right. The front panel dial shaft was therefore at right angles to the capacitor shaft and drove the capacitors either through a right angle worm drive or a cable and pulleys (Plate 9). This kept the traditional layout with its wide panel and small depth. The front-to-back arrangement came into its own much later when sets started to use electrical shielding (Chapter 10). The shielding allowed the tuned circuits for each stage to be placed closer together and still obtain a reasonable cabinet depth.

Also in 1924 the radio amateur, J. L. McLaughlin, built a three stage Neutrodyne which, of course, required four tuning capacitors and four tuning knobs all arranged in the traditional manner, side by side with the shafts running front to back (McLaughlin 1924). Being that the adjustment of four knobs all together is difficult, to say the least, he devised a one-knob system. He affixed a gear to each shaft and connected them with a rack (linear gear) (Fig. 7.1). This must have been the first time that "rack-and-pinion steering" was applied to radio!

Perhaps it was a coincidence, or McLaughlin had inside knowledge, but at the same time Carl Strube, a student of Hazeltine, was working at the Hazeltine Laboratory on the same idea (Wheeler and McDonald 1931). This design was used in the Thermiodyne TF5 and continued to be used in many of their later sets. The King Model 80 also used a similar scheme (Plate 19).

There were differences in the details of the systems. McLaughlin used short pieces of brass rack attached to a strong strip of bakelite which was spring loaded against the gears to avoid backlash. Strube used a long continuous brass rack (Douglas 1991) as did King. In order to help in electrical tracking McLaughlin mounted an extra trimmer capacitor with its own front panel knob above each ganged capacitor. Strube was able to afford a new capacitor design for the mass-produced Thermiodynes which neatly solved the trimmer problem. He made the capacitors with concentric shafts, the outer shaft operating the moveable plates as usual while the inner shaft, continuing to the rear of the unit, turned a single-plate trimmer capacitor (Plate 18). King used a lever and link to turn the stator of the antenna tuning capacitor through a small angle and thereby compensated for tracking error (Plate 21).

With the issuance of Chamberlain's one-shaft patent it didn't take long for engineers to invent other schemes besides the rack-and-pinion. One popular design used belts to couple the capacitors together. In the traditional orientation, left to right with shafts perpendicular to the panel, it was a simple enough proposition to put a pulley on each shaft and couple them with a belt. The details again varied among manufacturers. Some, like the Freshman 7F2 used flat brass belts (Radio News 1927). Others, like the Thompson "Minuet" used stranded steel wire (Rowe 1926). Usually the belts were anchored to the pulleys so that they wouldn't slip although some designs used belts that were designed to slip allowing each dial to be turned individually for fine adjustment. This idea was carried out in a more formal way in Dayfan's Model 5 which was provided with individual slipping clutches on each pulley (Radio News 1927).

Thus most of the belt driven sets still retained the traditional three tuning dials although they could be all driven

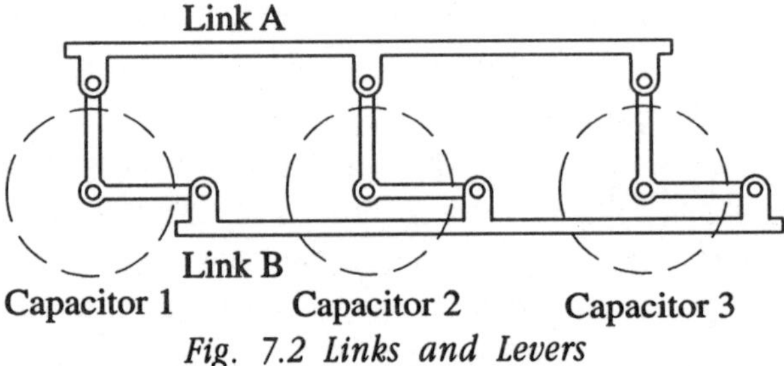

Fig. 7.2 Links and Levers

together with only one of the knobs. Whether these sets really qualified as "one-knob" or not they did simplify the tuning for the average listener.

The Grebe MU1 that was described in Chapter 5 provided a connection between its three tuning dials in a novel way. The method (Radio News 1926) takes advantage of the fact that the capacitors, although still in the "traditional" left-right configuration, had their axes rotated so that their shafts were vertical (Plates 15, 16 and 18). The tuning knobs were mounted on top of each shaft with their rims projecting through the front panel. This provided an additional sales "feature", that is, the dials could be rotated with the operator's thumb. More to the point, it allowed the designers to mount a sprocket on top of each dial and link them together with a beaded-link chain, similar to the chains still used on electric lamp switches today. A double sprocket was mounted on the center capacitor shaft to drive two chains, one driving the left and the other driving the right capacitor. The chains were given sufficient slack so that each knob could be independently adjusted through a small angle for fine tuning adjustment. The center dial could then be used to tune in a station and final tuning adjusted with the other two dials. Friction clutches or auxiliary trimmer capacitors were therefore unnecessary.

Another method for coupling the capacitor shafts, whether they were placed with their shafts horizontally or vertically, was by means of a linkage mechanism, Fig. 7.2. In 1926 both Music

Master (Radio News 1926) and Pfanstiel (Radio News 1926) were among the manufacturers using this method. Two levers at right angles on each shaft were connected to a common bar. Since the standard capacitors rotated a half turn from maximum to minimum a single lever and connecting bar would have, at some point in the rotation ended up in line (on "dead center") and the mechanism would have jammed. But with two levers, when one set of levers was parallel with the bar the other set provided the required leverage through the other bar.

By the end of 1926 the use of capacitors all mounted on a common shaft was a well used technique. Instead of mounting each capacitor separately and connecting them with short couplings the engineers designed them as one complete tuning unit. Two sets, by Perlesz and Ferguson used these tuning units with the capacitors arranged left to right on a single shaft.

These capacitor units were ruggedly constructed and were praised by Rowe when he wrote that the Perlesz set's

> details and ensemble may justly be termed works of mechanical art. In the opinion of old-timers, no less than of the novices, a receiver which operates satisfactorily with but one control is "a consummation devoutly to be wished": and when such a receiver embodies craftsmanship of such high order that every line of its appearance is pleasing, it deserves more than momentary attention. (Rowe 1926)

Not only were the capacitors mounted as one unit but at the right end the unit contained an integral worm gear drive for the tuning knob. At the other end a cylindrical dial was attached to the shaft and the settings were read through a decorated opening in the front panel.

The Ferguson unit comprised four ganged capacitors with the dial and worm drive placed in the center. In addition, a variocoupler for the antenna circuit was directly coupled to the left end of the shaft. The whole unit of cast aluminum was, as Radio Engineering April 1926 photo caption put it, "a real radio machine, designed to be put together quickly and to stay put forever."

In the 30's ganged capacitors were mass produced and became standard parts available to all radio manufacturers. The mass produced units may not have been as elegant as the Ferguson unit but millions of people used them for simple "one-knob" tuning.

Chapter 8

The Electronics of One-Knob Control

Even the best mechanical "one-knob" arrangement wouldn't have been acceptable if it had seriously spoiled the set's electrical performance. Even when all the tuning capacitors were made precisely the same and accurately driven together there still remained the electrical problem of making each RF transformer tune together. This is the "tracking" problem that the engineers had to solve if the one-knob set was to be practical. In many designs the tracking was not exact and extra controls were provided so that the operator could compensate for the tracking error. Usually these controls were either additional variable capacitors connected across each of the main tuning capacitors or, as described in the last chapter, the mechanical ganging system provided the necessary small individual motion of the capacitors. But a true one-knob set had to do without auxiliary knobs and required some electrical innovation.

The largest errors in tracking were produced by the action of the antenna on the first RF transformer. Manufacturers had no control over the type of antenna the customer might use. Antennas of different lengths would change the tuning of the antenna circuit. This problem was further aggravated because the first RF transformer was designed to closely couple the antenna to the secondary. This close coupling provided a strong signal and made the sets as sensitive as possible even at the expense of antenna interaction. So that even when all subsequent stages tracked well a separate antenna variable capacitor had to be provided.

The cost of an extra variable capacitor could be avoided by an ingenious mechanical scheme. In this design the stator of the antenna tuning capacitor wasn't fixedly mounted but allowed to turn. A linkage arrangement from the tuning knob shaft rocked the stator through a few degrees effectively adjusting the tuning capacitor over a small range. This arrangement was used in the King Model 80 in 1927 (Plate 21).

True one-knob control without trimming the antenna circuit could be achieved if the manufacturer supplied a built-in loop antenna. The design of the loop was under the control of the manufacturer and made it unnecessary for the customer to supply an external antenna. The loop usually was used in place of the first RF transformer. It was tuned by the first tuning capacitor and, when it was made to have the same inductance as the tuned coils in subsequent stages, good tracking could be achieved. Many loops were small enough to be built into the radio cabinet. When they did not provide satisfactory reception a large external loop up to 3 or 4 feet square was used. Many loops were designed to fold up much like an umbrella so that they could be stored within the set. In many sets the first RF transformer and trimmer capacitor were retained so that the loop could be disconnected and a long antenna used instead. Today every AM radio uses a small "loop" or coil wound around a ferrite magnetic core to provide the same reception as the old much larger loops.

Another method for eliminating the antenna trimmer, but at the expense of performance, was to use an untuned antenna circuit. The untuned circuit allowed amplification of all frequencies and did not give the boost to the selected station signal that a tuned circuit would have given. The resulting loss of sensitivity required an extra RF amplifier stage to restore the set's performance. The Freshman 7F2 used this design with a high inductance antenna coil instead of a tuned RF transformer in the grid of the first stage. This stage was followed by two more RF amplifiers which gave a total of three RF stages and three RF transformers. The three RF transformers were tuned by three ganged capacitors. In order to keep strong signals from overloading the amplifiers Freshman provided a front panel switch that could be used to connect a resistor across the antenna coil and reduce the input signal. So, although the set did not have a separate antenna trimmer capacitor it still, at times, required the operation of an additional switch.

When more than two stages of tuned radio frequency amplification was attempted good tracking became even more difficult. As described in the last chapter, the amateur McLaughlin coupled the four knobs of his three stage set with a rack-and-pinion drive. Possibly because of all the problems he jokingly called his set the "Super Calamityplex".

His mechanical solution was all well and good but there still remained the electrical tracking problem. He admits that

> The adjustments of the condensers so that they all work in synchronism was the difficult part of building this receiver—but since all the tuning circuits were in balance no serious difficulty was encountered. (McLaughlin 1924)

But it seems that there was a "serious difficulty", the circuits didn't really track, and trimmers were required as McLaughlin adds that

> As an aid in fine tuning three small variable condensers, of three plates each, are used. They need not be adjusted for each change in wavelength but are intended to put the tuning

> system into balance, after which it is controlled by a single dial. One of these small condensers is connected across the first, second and fourth main condenser; there is none across the third main condenser.

The trimmer across the first capacitor was definitely needed to take care of the antenna tuning. The trimmer capacitors across the second and fourth stages were required to solve a different problem. McLaughlin's set was built on a breadboard with a bakelite panel like all the other sets of the early 20's. No metal shielding of any kind was used. The first and third RF transformers had their axes aligned horizontally, the second and fourth aligned vertically. Thus the first and third, as well as the second and fourth, had their axes parallel and although spaced widely apart a significant amount of coupling must have existed between them. Many other stray capacitances between the coils, tubes and capacitors added to the electromagnetic interaction. All these stray couplings provided unwanted signal paths that degraded the performance of the RF amplifiers. As Dreyer and Manson (1926) point out they particularly upset McLaughlin's Neutrodyne circuits and made neutralization difficult. It's likely that McLaughlin ran into this problem as he says that the "the second stage is only partially neutralized. This allows the middle tube to oscillate feebly and helps the strength of weak signals". It perhaps was a wonder that the set tracked at all without all stages neutralized. It is also significant to note that, even with three stages some regeneration was accepted in order to obtain the required sensitivity. Radio engineers had to fight hard in those days to obtain satisfactory amplification with the old triodes!

Before leaving this discussion of one-knob radios it is worthwhile to give special mention to a one-knob set with a unique configuration, the Thompson "Minuet" made in August 1925. In addition to its one-knob control using a steel belt it had many other features that made it ahead of its time. The company did not exaggerate too much when they said in their advertisement that their set is "Unlike Any Model On the Market". As modern industrial engineers would say, "It's all in the packaging". Electrically it used the usual circuit of its era: Two stages of radio frequency amplification, a non-regenerative detector and two stages of audio frequency amplification. It was the appearance and means of control that set the Minuet apart.

Imagine a horseshoe standing upright on its two legs, the circular part of the shoe some 18 inches in diameter with a cone loudspeaker mounted in the center. This is the front view of this unique set and brings to mind the cathedral and tombstone table top radios of the 30's. A circular dial around the upper half of the round loudspeaker surrounds a semicircular slot from which protrudes a small handle and pointer. Moving the handle through the semi-circular arc tunes the receiver. The handle is pivoted from a point directly behind the center of the speaker and, through a series of pulleys driven by wire cord, turns all three tuning capacitors at one time. Thus "one knob" tuning was accomplished in an unusual way, the control and dial occupying very little panel space around the upper periphery of the loudspeaker.

Below the loudspeaker along a horizontal line were four knobs and the on-off switch. The first knob, the "volume control" controlled a filament rheostat for the two radio frequency stages. The second knob controlled a separate filament rheostat for the filaments of the other three tubes. The other two knobs controlled, through levers, the stators of two of the tuning capacitors. The stators were mounted so that they may also be turned through a small angle and tune the circuits independently to make up for the poor tracking of the circuits. As Radio News' G.C. Rowe puts it:

> This elaborate lever system may, on first thought, seem unnecessary and complicated in operation, but this is entirely false. This variation of the stator plates is for the equalization

> of any irregularities in the radio frequency transformer windings. When these plates are once adjusted there is no need for disturbing them until the set is used under other conditions.
> (Rowe 1926)

Rowe is not only trying not to put too much blame on the coil winders but also puts the best "spin" on the tuning procedure. By "other conditions" he means that when the set is tuned from a station to another one far away on the dial the operator must, most often, readjust these "compensating tuning control knobs".

The "Minuet" used tubes designed for dry battery operation which dispensed the need for the storage battery. All batteries were self-contained within the set.

In the same article Rowe writes eloquently about the radio scene of his day and was inspired by the new look of this set. His remarks give us some insight into the status of broadcast radio receivers in 1926:

> Within the last few years science in general has stridden forward with steps that would do credit to the "seven-league boots" that fascinated us as children. Thanks to the tireless efforts of many patient workers those same tales that were so real and yet improbable, are things of reality today. It is needless to reiterate what these manifold wonders are, because daily we use some of them and think nothing of it. However, because of its recent popularity we are more inclined to appreciate the advances that have been made in radio.
>
> As recently as two years ago little thought was given by the radio set constructor to the aesthetic side, and if a set had more controls than were necessary to operate a battleship, it made little difference as long as some favorite station could be picked up with fair volume. . .However, since radio has become so firmly established as a national pastime, the demand for more decorative cabinets and receivers that could be operated by anyone from grandma down to the

> baby has been answered by the multitude of attractive receivers that have appeared on the market within the last few months. (Rowe 1926)

These "manifold wonders" from those "tireless and patient workers" were within a decade to change radio from a "national pastime" to a national necessity when no one could do without a radio in their home.

Chapter 9

Special TRF's

The "Lossers" by and large accounted for a large proportion of the 1920's radios along with their more elegant brothers, the Neutrodynes. However there were also many "poor relations" in which engineers used ingenious ways to neutralize their old nemesis, the triode. One of these was the bridge neutralizing circuit based on the simple Wheatstone bridge.

The familiar direct current Wheatstone bridge circuit (Fig. 9.1a) consists of four resistors, a battery and a voltmeter. The battery voltage is applied to the top and bottom (points 1 and 4) of the bridge and the voltmeter is connected between the left and right (points 2 and 3). When the values of R1 and R2 are in the same proportion as R3 and R4, no voltage appears across the voltmeter and the bridge is "balanced". This result is easily seen to be true by considering the current through the two branches. The battery current splits between the two arms. According to Ohms's law the voltages across R1 and R2 are in proportion to the resistor values. Similarly the voltages across R3 and R4 are proportional to their values. Since the resistors in the two arms are equally proportioned, so to are the voltages and the same proportion of the battery voltage appears across R1 and R3 or across R2 and R4. Thus the voltage from point 4 to points 2 and 3 are equal and no voltage appears across the voltmeter.

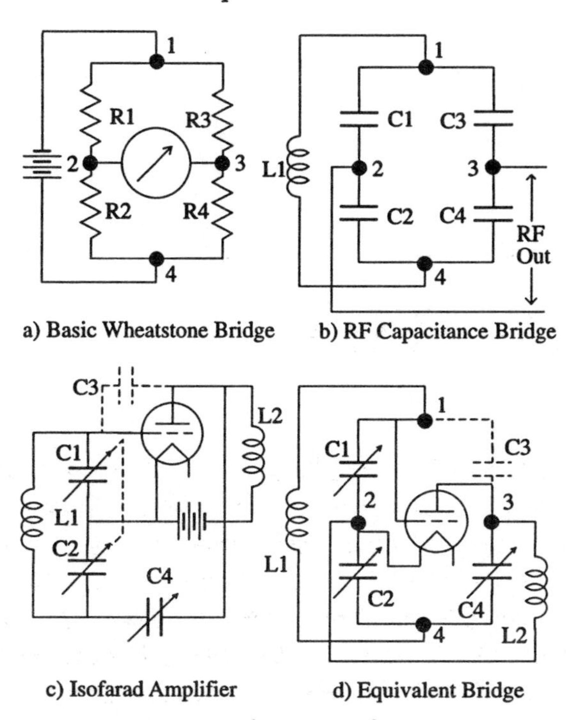

Fig. 9.1 Bridge Neutralization 1

An important feature of the bridge circuit is that, when it is balanced, the battery and voltmeter can be interchanged and the balance remains undisturbed. Thus any voltage across points 1 and 4 does not appear across points 2 and 3 and neither does voltage across 2 and 3 appear across 1 and 4.

The Wheatstone bridge can be made to work at radio frequencies by replacing the battery with an RF signal source and the voltmeter with a detector. The arms of the bridge may then consist of capacitors, inductors, resistors or combinations of all three. An RF bridge using capacitors only, Fig. 9.1b, can supply a useful function in radio receiver design. The input signal to the bridge is supplied by the secondary, L1, of an RF transformer and the voltage across points 2 and 3 is sent to the next amplifier stage. As in the case of the Wheatstone bridge,

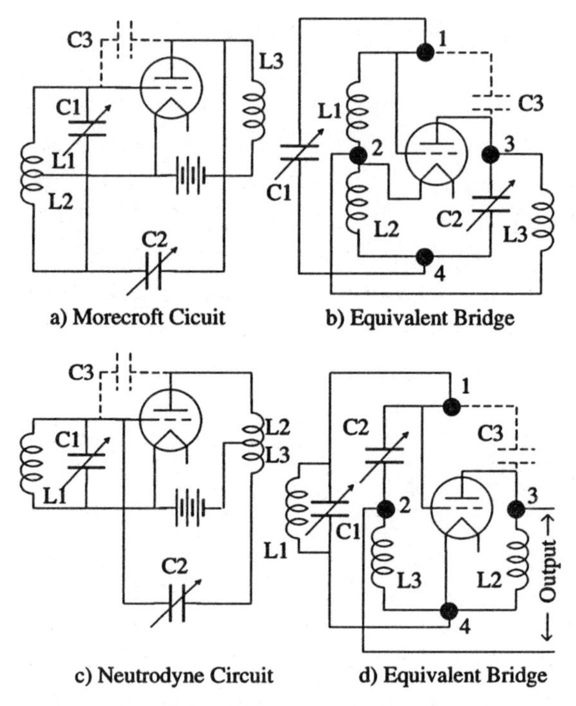

Fig. 9.2 Bridge Neutralization 2

when the capacitors C1 and C2 are in the same proportion as C3 and C4 the bridge is balanced and the input signal will not appear at the output. Conversely, as in the Wheatstone bridge, the output voltage will not affect the input so that the input and output do not interact.

The capacitance bridge was used by Byron P. Minnium, Radio Engineer of the Walbert Mfg. Co., to neutralize the grid-plate capacitance of the RF amplifiers of his receiver. In his circuit, Fig. 9.1c, which he called the "Isofarad", the capacitors C1 and C2 are two identical sections of a two-gang tuning capacitor which tune the secondary of the input RF transformer, L1. These capacitors always remain in the same proportion, equal to each other, as the circuit is tuned. The stage is neutralized by the capacitor, C4, which counteracts the effect of the tube's grid-plate capacitance, C3. The Isofarad circuit can be redrawn, Fig. 9.1d, to correspond to the capacitance bridge, Fig. 9.1b. The capacitors, C1 and C2, become the two sections of the tuning capacitor and C3 and C4 become the grid-plate and neutralizing capacitors. When the neutralizing capacitor, C4, is adjusted to equal the tube capacitance, C3, the bridge is balanced. The output signal then cannot appear on the input to cause the amplifier to oscillate and the stage is neutralized.

The Isofarad, like the Neutrodyne, is an efficient means of neutralization unlike the losser circuits (Chapter 6). Minnium points out that his circuit

> eliminates at its source the chief obstacle to efficient radio frequency amplification. . .This is, of course, contrary to the usual custom of preventing self-oscillation by the addition of resistance in the secondary circuits, the use of very few primary turns . . . and a corresponding reduction in coupling between primary and secondary. Such schemes are definitely limited to an approximate approach to the point of oscillation and make very little use of pure repeater action in amplification. (Minnium 1925)

Although the Isofarad is an efficient means of neutralization it has the disadvantage that the tuning capacitors reduce the signal at the grid of the tube to only half of that delivered by the input transformer. The gain of the stage is therefore only half of a similar Neutrodyne stage.

Like the Isofarad, other neutralizing circuits can also be described by using the Wheatstone bridge analogy. The Neutrodyne, Chapter 4, was described in terms of the principles of feedback. It was shown that the operation of the neutralizing capacitor was to couple a voltage from the output to the input that opposed the effect of the tube's capacity. This feedback principle was used by Morecroft (1927) to describe the operation of a neutralization circuit, Fig. 9.2a, which he calls "neutralization through the grid circuit". The neutralizing capacitor, C2, feeds the signal from the plate to the bottom portion, L2, of the grid coil. This signal couples an equal and opposite voltage to the top portion, L1, of the grid coil which cancels the signal fed back through the grid-plate capacitance, C3.

As with the Isofarad, Morecroft's circuit can be explained by redrawing it as the equivalent bridge, Fig. 9.2b. The bridge contains, on the left, the two inductances, L1 and L2, of the tapped grid coil and, on the right, the grid-plate capacitance, C3, and the neutralizing capacitor, C2. The bridge is balanced when L1 equals L2 and C2 equals C3 and, like the Isofarad, the grid-plate capacitance is neutralized.

Similarly, the Neutrodyne circuit, Fig. 9.2c, can be redrawn as the bridge circuit, Fig. 9.2d. The equivalent bridge contains, on the left, the neutralizing capacitor, C2, and half the plate coil inductance, L3, and, on the right, the grid-plate capacitance, C3, and the other half of the plate coil, L2. The bridge is balanced when C2 and C3 are in inverse proportion to L2 and L3. Equal voltages are induced in the secondary of the output transformer from L2 and L3 which is equivalent to the voltage across points 2 and 3.

Thus all these neutralizing schemes can be shown to be balanced bridges. Whether they are described as a bridge or a feedback circuit their basic operating principle is the same. Minnium went to great lengths to describe his circuit as a novel bridge neutralizing method but it worked on much the same principles as the Neutrodyne and Morecroft's grid circuit. He manufactured Isofarad sets from 1925 to 1927 and then went out of business. He might well have been enjoined to stop production if RCA had finally realized that he was really using a modification of their patented Hazeltine circuit.

Walbert shielded the coils in his sets by placing them in aluminum cans, an earlier forerunner of shielded sets (Chapter 10). He used three RF stages requiring four tuning capacitors with their four tuning knobs. Unlike other manufacturers of the same period he did not attempt to gang them together. In an article in Radio News Griffin believes that the great selectivity of Walbert's set could not be obtained without four separate controls. He puts his position as follows:

> Single control is a commercial impossibility, if each of the several circuits operated by one knob tunes sharply. . . . In consequence multiple control is used; and the circuit is so designed that it will cut out very powerful locals and bring in distant stations through bad interference. The receiver is adjusted so that the dials read approximately the same at all wavelengths. . . As human beings are supplied with two hands, there is little excuse for not using them.
>
> (Griffin 1926)

There are always those who would cling to the old ways!

At about the same time Minnium invented his bridge stabilization circuit Harry A. Bremer developed another variation of the Neutrodyne, the "Counterphase" circuit. The simplified schematic, Fig. 9.3, shows the three RF stages that preceded the detector. Each stage was neutralized by the neutralizing capacitors, C1, C2 and C3, in what at first seems like the Neutrodyne circuit. But in the Counterphase the capacitors were connected, like the Morecroft circuit, to a tap on the RF transformer secondary. However, unlike the Morecroft circuit and like the Neutrodyne, neutralizing voltage was obtained from an auxiliary coil in the plate circuit.

Perhaps the most unique part of Bremer's circuit was the addition of the variable resistor, R2, in series with the neutralizing circuits of stages two and three. The operation of this control is described in a Radio News article as follows:

> Each stage is easily adjusted so as to prevent oscillation at any frequency by varying the series resistance. Decreasing the series resistance increases the tendency to oscillate, thus

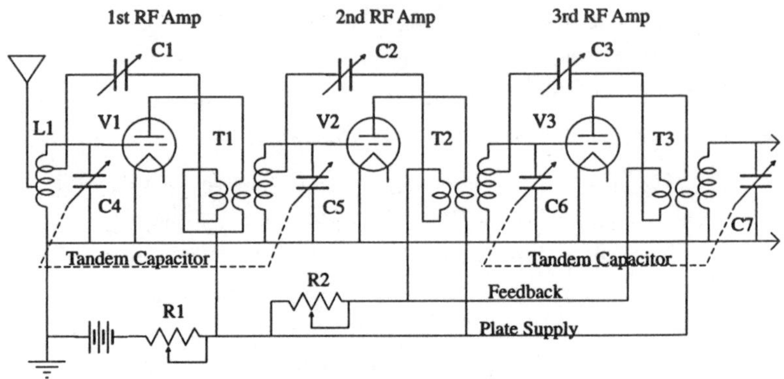

Fig. 9.3 Counterphase

> governing the amount of reverse phase energy necessary to suppress oscillations at high frequencies. (Carlton 1925)

It would seem from this description that the neutralization was not effective at all frequencies due, no doubt, to stray coupling between stages. The RF transformers used toroidal windings to minimize electromagnetic coupling but still any remaining inductive or capacitive coupling could have seriously affected the neutralization. The resistor, R1, which adjusted the plate voltage was mechanically combined with R2 to produce a combined volume and regeneration control. The four tuning capacitors of the three RF amplifiers were ganged in pairs so that "only" two tuning knobs were required. Even then each "tandem" capacitor had to have a separate trimmer (not shown) to keep the two circuits in tune and one-knob tuning would have been difficult to achieve. But, like the Walbert set, two knobs were acceptable at that time as "no normal, two-handed person wants a radio set with only one dial to turn" (Carlton 1925).

In describing the early Freshman sets that used the losser method of stabilization, Mesa (1927) found it not entirely satisfactory and designed a new neutralizing circuit. His "Equaphase" circuit, Fig. 9.4a, consists simply of the two resistors, R1 and R2, and the capacitor, C1, connected across the primary of the RF transformer, L1. Unlike the circuits already

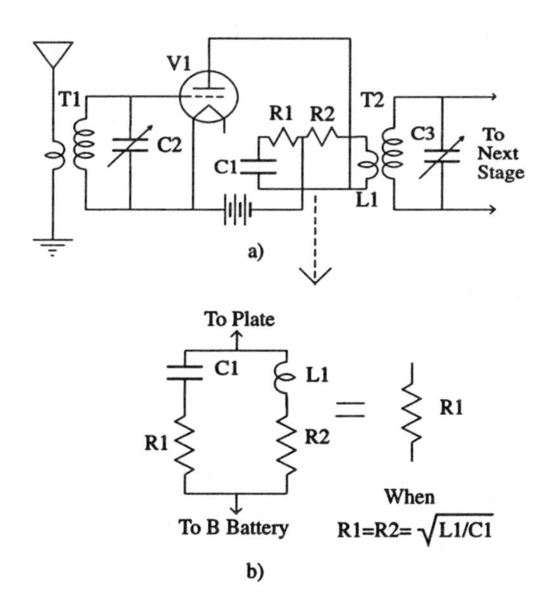

Fig. 9.4 Equaphase

discussed no feedback is used to neutralize the effect of the tube's grid-plate capacitance. The four components in the plate circuit comprise a tuned circuit, Fig. 9.4b, having the resistors, R1 and R2, in series with both the inductor and the capacitor. The resistor R2 also includes the inherent resistance of the windings of inductor, L1. For certain values of the components this tuned circuit has a peculiar characteristic. If R1 and R2 are equal and also equal to the square root of L1/C1 the circuit acts as though it were a resistor with a resistance equal to R1 or R2 (Morecroft 1927 p.92). The tuned circuit no longer tunes but acts at all frequencies as a resistor.

The Counterphase circuit achieved this condition by the adjustment of R1 and R2 to produce a purely resistive circuit in the plate. Mesa describes the operation as follows:

> Therefore, the plate circuit of the previous amplifier has no inductive reactance in it and so the tube cannot oscillate,

> provided the values of inductance, capacity, and resistance are properly adjusted. (Mesa 1927)

It is difficult, however, to believe that this was all there was to it. The losses introduced by the effective resistance, R1, like the resistors in the "losser" circuits, may well have reduced the amplifier gain sufficiently to prevent oscillations. However the circuit worked well enough to be put into production and in 1928 Freshman made both battery operated and AC operated sets using the Equaphase circuit.

All the sets so far have used circuits to prevent oscillations in the RF amplifiers but in 1925 Kennedy purposely put regeneration back into the RF amplifiers of their Model XXX type 435 (Griffin 1926). Instead of using a regenerative detector Kennedy used Armstrong's regeneration principle to increase the gain of an RF amplifier. His set used two RF stages followed by the normal non-regenerative grid-leak detector. The second RF stage coupled energy back to its grid circuit with an adjustable tickler coil. The rotating tickler was mechanically ganged with a filament rheostat and acted as a volume control. Highest gain was provided when the filament voltage was maximum and the regeneration was increased to the point of oscillation. The volume control was arranged to first reduce the regeneration and then reduce the filament voltage which effectively reduced the sensitivity for the reception of strong stations.

Alexanderson's patent for the TRF was held by the RCA-GE-Westinghouse consortium and independent TRF manufacturers had to pay for a patent license. The cost of a patent license was a great incentive for engineers to innovate ways to find an alternative. One way to get around Alexanderson's patent was to use an UN-tuned RF Amplifier. A true untuned amplifier would amplify all signals without regard to frequency. A TRF receiver relies on tuned RF transformers to provide its selectivity. How could an untuned set provide the necessary selectivity? Of course the antenna circuit could still be tuned without

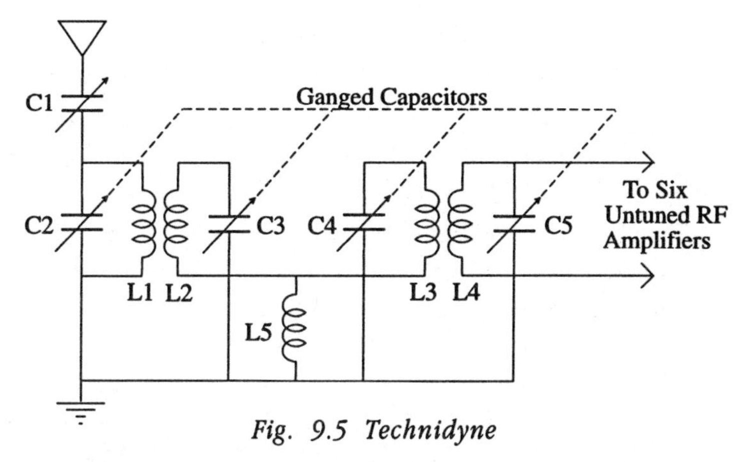

Fig. 9.5 Technidyne

infringing on Alexanderson's patent. But a single tuned circuit could not give the selectivity of a two-stage TRF with its three tuned circuits. Additional tuned circuits were required.

Lester L. Jones was Hazeltine's colleague and helped develop the Neutrodyne circuit. But Jones didn't receive any monetary gain from Hazeltine's patent so he went off on his own to design a set, the "Technidyne", that wouldn't infringe on Hazeltine's patent. Jones' amplifiers used untuned RF transformers which, because of their low gain per stage, didn't need neutralization. Thus, his amplifiers didn't infringe on Hazeltine's patent and, because they were untuned, they also didn't infringe on Alexanderson's TRF patent. Jones provided all the selectivity by means of a special antenna tuner, Fig. 9.5, that contains four tuned circuits. The capacitors of the four circuits, C2, C3, C4 and C5, were ganged to provide one-knob control. The antenna was tuned by an additional variable capacitor, C1, which also connected the received signal to the first tuned circuit, C2 and L1. The signal was then inductively coupled to L2 which formed the second tuned circuit with C3. A common coil, L5, transferred the signal to the third tuned circuit, C4 and L3. Finally, the signal was coupled from L3 to L4 which formed the fourth tuned circuit with C5. After passing from the antenna through this tuner the signal was amplified by multi-stage untuned amplifiers using high inductance closely wound untuned coils (Plate 7). The gain of each stage was much lower than that of tuned amplifiers so that as many as six stages were required to obtain the desired sensitivity.

Jones patented the Technidyne and sold licenses to the manufacturers Sparks-Withington and AC Dayton. These sets were manufactured in 1927 and 1928 under the brand names "Equasonne" and "Navigator". Although more RF amplifier tubes were used than in an ordinary TRF it didn't matter as much as in earlier sets. By the late 20's, sets operated from AC power, not batteries, and filament battery drain was unimportant. Jones thus achieved his goal and the companies paid license fees to him and not to his old colleague Hazeltine or to the RCA/GE consortium.

At times, the Technidyne's four tuned circuits or the typical TRF's three tuned circuits did not provide adequate selectivity to prevent a strong local station from interfering with weak signals. In a technical article devoted to the analysis of radio interference, Goldsmith (1926) gives a circuit of a "wavetrap". The term "wavetrap" had its start in the 20's and, even today, wavetraps are made to avoid television interference by local TV stations. The basic idea is to attenuate the unwanted signal between the antenna and the receiver's antenna terminals. The Goldsmith wavetrap, Fig. 9.6, consists of a tuned circuit, L1 and C1, which is tuned to the local station's frequency. The circuit presents a high resistance to the local signal and, together with capacitor, C2, attenuates the powerful signal while leaving other signals unaffected.

The Bremer-Tully 8-12 had a wavetrap or "rejector" circuit permanently incorporated in the set (Radio News 1927c). The variable capacitor that tuned the rejector circuit was ganged to the other main tuning capacitors. A front panel switch was provided to turn the wavetrap on and off as interference from

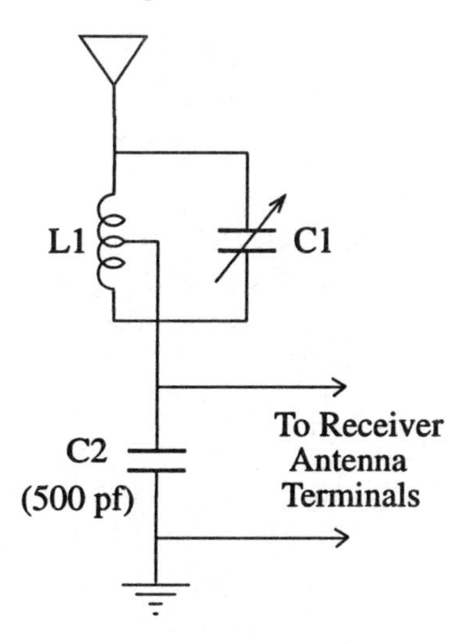

Fig. 9.6 Wavetrap

adjacent stations required. The wavetrap was followed by a three stage TRF amplifier that, in a break from the breadboard tradition, was completely shielded.

Bridge neutralization circuits, Isofarads, Counterphases, Equaphases, Technidynes, regenerative RF stages and wavetraps are all examples of unusual circuits during the era of the TRF in the mid 20's. These circuits were used by a small group of independent manufacturers and they illustrate the ingenuity of engineers in their competition with the large companies, notably RCA, Atwater-Kent, and Mohawk. They served their purpose at the time and enabled their manufacturers to offer a unique feature to their customers and, by avoiding royalties, to provide radios at a reasonable price.

1. Variometer (General Radio). This compact unit has the stationary coil wound inside the outer bakelite form. Connections to the rotary coil are made through stub shafts at each end of the unit.

2. Variometer (Manufacturer Unknown). Inside the two hemispherical coils is the rotating coil. The outer coil is tapped and connected to the rotary switch. The whole assembly is designed to be mounted on the front panel.

3. Variocoupler (Manufacturer Unknown). The rotating coil is visible inside the outer coil which has six taps. The taps are selected by the switch at the bottom of the photo. Contact is made to the inner coil by the springs on the left and right of the coil which bear against metal rings on the shaft.

4. Radiola III (RCA). The space behind the panel of this two tube set is almost completely taken up by the dual variocoupler. One of the two rotating coils can be seen at the right end of the coil assembly. A similar coil is mounted inside the other end. The AF transformer and filament rheostat are visible behind the variocoupler.

5. Regenaformer Coil (National). These two views of the second RF transformer show the coils and capacitors built by National to Browning-Drake specifications. The top photo shows the rotating tickler coil and the lower photo shows the "slot" coil mounted inside the secondary. The capacitor plates are curved to provide straight-line frequency tuning.

6. Doughnut Coil (Thorola). This coil is self supporting and its toroidal shape minimized coupling. The primary winding is wound between the secondary turns at one end.

7. Untuned RF Transformer and AF Transformer (Federal and King). The RF transformer covers the broadcast band and is sealed within a bakelite tube. The laminated iron core of the AF transformer has an air gap, visible on the lower right, which prevents the core from magnetic saturation by the DC plate current.

8. Tuning Capacitor with trimmer (U.S. Tool Co.). This multiplate capacitor is adjusted by the outside large knob. The inner knob adjusts the single plate unit in the rear of the unit by means of a concentric shaft. The plates are semicircular providing a closely linear variation of capacity with rotation.

9. Tuning Mechanism (National). The two tuning capacitors are coupled by the drum dial which is driven from the front panel control by a heavy cord. Note the shape of the straight-line frequency capacitor plates.

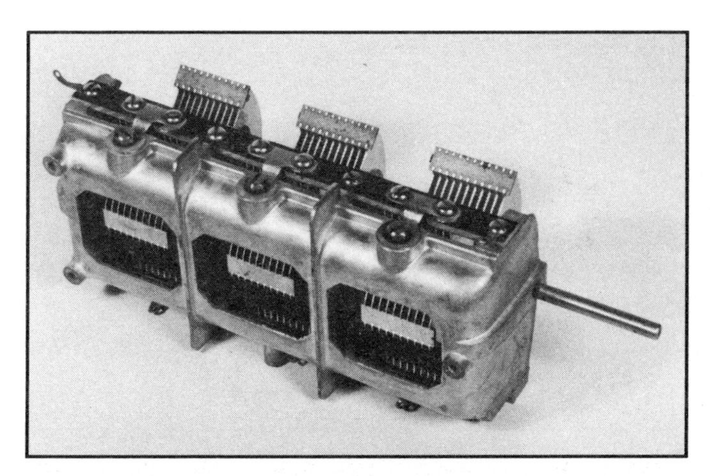

10. Ganged Tuning Capacitor (Manufacturer Unknown). This high production unit has a rugged die cast frame. The three sets of movable plates are pinned to a single shaft. The three large screws on top of the unit adjust three mica compression trimmer capacitors.

11. "Masterpiece" Tuning Unit (Freshman). This compact unit contains a honeycomb self supporting coil, a tuning capacitor, a dial with an internal gear and a knob and drive gear. Note how the coil is mounted close to the capacitor frame to introduce losses which help stabilize the set's RF amplifiers.

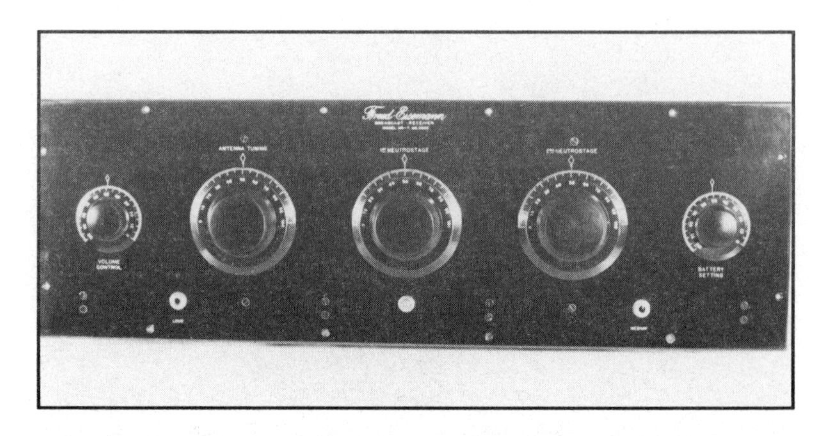

12. Freed-Eisemann NR-6. This handsome bakelite panel, 28 ins long and 9 ins high, is typical of early 20's radios. The three tuning dials are simply calibrated 0 to 100. The end knobs control the filament current. The headphone jacks and the filament switch are at the bottom of the panel.

13. RF Transformers (Freed-Eisemann). The three transformer coils are mounted with their axis at 57 degrees calculated by Hazeltine to minimize the coupling between them. Tuning capacitors are mounted on the front panel directly behind the transformers. Note the brass-rod neutralizing capacitors on the horizontal bakelite bar.

14. Grid Leak Detector (Freed-Eisemann). The AF transformer and filament rheostat are to the left of the tube. To the right is the third RF transformer and tuning capacitor. The grid leak resistor is below the bakelite mounting board and the capacitor is above behind the tube base. Note the RF transformer primary wound on a smaller form inside the secondary.

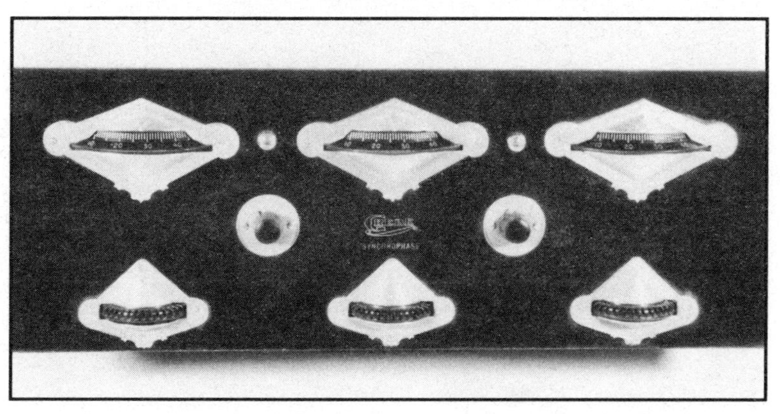

15. Grebe MU-1 Front Panel. This set is easy to recognize with its three thumb operated dials with small vernier dials below. The small knobs are labeled "tone color" and "volume". The dials are calibrated simply 0 to 100.

16. Grebe MU-1 Rear View. This view shows the chain drive between the thumb wheel dials, the binocular coils and an enclosed AF transformer. The layout is "traditional" with the RF amplifiers progressing left to right as seen from the front of the set.

17. Binocular Coils (Grebe MU-1). This close-up, with the tubes removed, shows the three coils wound on double cylinders so as to minimize the coupling between them. Immediately above the coils is a slide switch for changing frequency bands. This switch was operated when the dials were turned to their end stops.

18. Chain Drive (Grebe MU-1). This closeup, with the tube removed, shows the chain connecting the single sprocket dial in the foreground to the double sprocket on the center dial in the background. The vernier control knob below the tuning capacitor drives the main dial through a vertical shaft and friction drive. Note the mica compression trimmer capacitor in front of the tuning capacitor.

19. King 80 Sub Panel. The front panel has been removed revealing the rack and pinion drive connecting the three tuning capacitors. The center capacitor shaft carries the circular dial which is friction driven from the tuning knob shaft. The dial is calibrated in meters (of wavelength) as well as the traditional 0 to 100.

20. King 80 Rear View. The RF transformers are mounted with their axes at right angles. The first and second RF amplifier tubes are near the front panel between the tuning capacitors. The detector tube is on the left (in the photo) and the three AF amplifier tubes follow to the right.

21. Tuning Capacitor (King 80). The close-up shows the rear of the antenna tuning capacitor. The lever and link that rotates the stator under control of a front panel knob is clearly visible. This adjustment was necessary for proper tracking in this one-knob set.

22. RF Transformers (Atwater-Kent 20). The three transformers are mounted at right angles to each other to minimize coupling between them. Note the elegant molded ends on the tuning capacitors and the molded tube bases.

23. AF Amplifier (Atwater Kent 20). The three tubes mounted on the perfectly molded base are, from right to left, the detector and the first and second AF amplifiers. The grid leak is visible on the right. The two AF transformers mounted in black cans are directly behind the tubes.

24. Superheterodyne (Remler). This set was assembled by the author using components that were originally supplied in kits sold by Remler. The set uses a loop antenna (not shown). The coil on the right is the local oscillator coil tuned by the rightmost capacitor. The other tuning capacitor tunes the loop. The first tube on the right is the local oscillator followed by the 1st detector, three IF stages, and at the extreme left the 2nd detector and two AF amplifiers. The four IF transformers in bakelite cases are in front of the IF amplifiers. The IF amplifier tube sockets are raised to the level of the tops of the IF transformers in order to keep the leads short.

Chapter 10

Shielded TRF's

Radio frequency amplifiers were greatly improved with the introduction of metal shielding in radio construction. Previously the electric and magnetic coupling between stages was reduced to manageable proportions by placing the RF transformers far apart and properly aligning their axes to minimize interaction. As described in Chapter 5 the three transformers of a two-stage amplifier could be aligned along the three space coordinates while a three stage amplifier could take advantage of Hazeltine's fifty-five degree angle, Fig. 5.2. The next step and one that seems so obvious to today's engineers was to put each transformer inside a metal can. When the can was not too close fitting and made of a good conductor, like aluminum, then the coils were not adversely effected as they were in the "losser" designs (Chapter 6).

Shielding the coils had a further beneficial effect. A TRF set relied on all of its tuned circuits to separate a weak station whose frequency was close to a strong local station. If the unshielded coil in the last stage happened to pick up the strong station the selectivity of that transformer alone would not be sufficient to separate the two stations. The overall selectivity could only be obtained by passing the signal through the tuned circuits of every stage. It was defeated if a powerful station "jumped over" to the last stage.

Walbert placed shielded cans around the RF transformers in his bridge neutralized three stage set (Chapter 9). Griffin reported in Radio News that

> the coils are shielded in metal containers which remove the last possibility of feed-back and consequent oscillations. . . The relation of the coil to the shielding is such that absorption losses are trifling. (Griffin 1926)

It was obviously a great improvement at the time but Griffin was too optimistic when he refers to the "last possibility of feed-back". The tuning capacitors, the wiring and other circuit parts were still unshielded and were a possible source of interstage coupling, and for best performance they too must be shielded from each other. This shielding was properly accomplished by using a metal chassis with shielded partitions.

One of the first commercial radios to use this method for complete shielding was the Stromberg-Carlson Model 601. In their article on this set, the engineers responsible for this project, John F. Dreyer, Jr. of Hazeltine and Ray H. Manson of Stromberg Carlson first discuss the development of the Neutrodyne circuit and the problems affecting that circuit in unshielded sets. They note that even when unshielded coils were mounted to minimize the inductive coupling the stray capacitance still existed. This capacitance interfered with proper neutralization and they pointed out that

> Capacity between non-adjacent stages (between the first and the last in a three-circuit unshielded receiver) may result in appreciable regeneration and in oscillation when the total radio-frequency amplification is increased beyond a certain point. Many Neutrodyne receivers are supplied with a third neutralizing condenser which neutralizes this over-all capacity. These receivers may obtain a somewhat higher degree of amplification than receivers which are not so supplied.
> (Dreyer and Manson 1926)

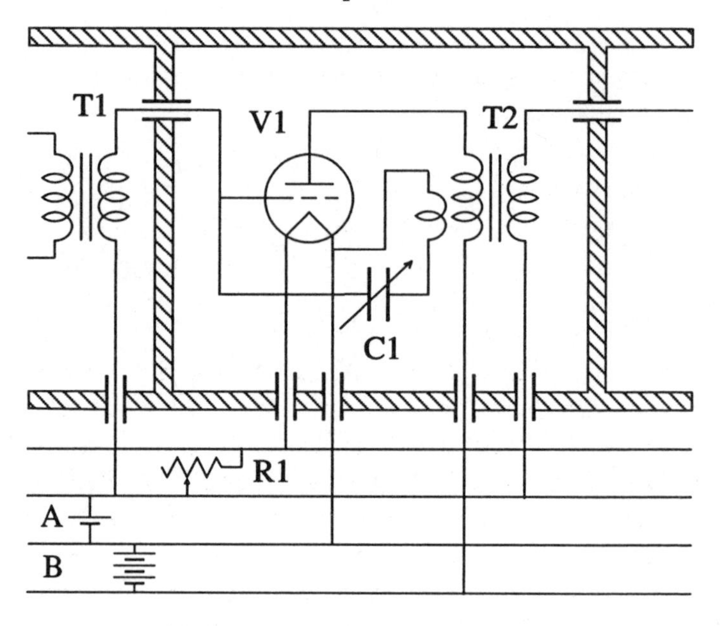

a) Hazeltine (1919)

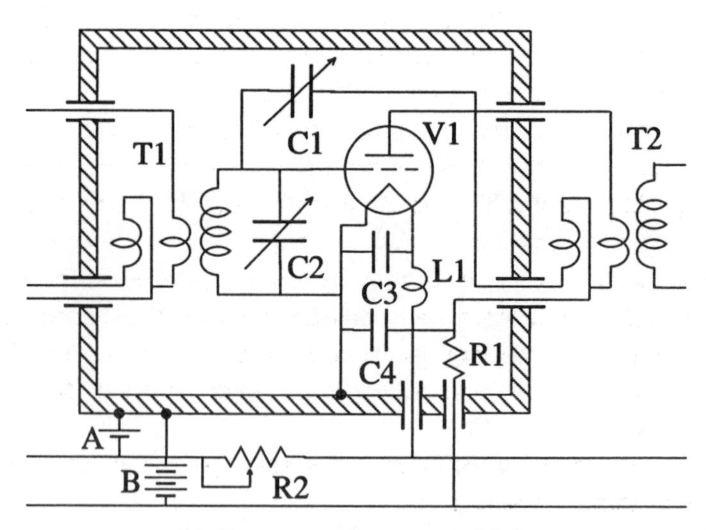

b) Dreyer - Manson (1926)

Fig. 10.1 Shield Neutrodyne

When more stages are added the capacitive couplings become more complicated and with three RF stages a total of six neutralizing capacitors might have been required. This led them to the conclusion that "if three or more stages are required, complete metallic shielding must be resorted to."

In their article Dreyer and Manson showed Hazeltine's schematic of a shielded amplifier that he had proposed as early as 1919. It reveals the fact that the principles of shielding were not new in the early 20's but their use had to overcome the inertia of tradition and the economics of radio set manufacture.

Hazeltine's shielding arrangement, Fig. 10.1a, shows the tube, V1, the transformer, T2, and the neutralizing capacitor, C1, contained in a metal compartment. Connections were made to the previous and later stages through small holes in the shield. The transformers were untuned and used an iron core. Although not stated by Dreyer and Manson it's presumed that the amplifier was proposed for high-frequency amplification made possible by the Neutrodyne circuit. Dreyer and Manson went one step farther than Hazeltine and placed each stage of their receiver in a separate can so that there were double metal walls between each stage. Each can, Fig. 10.1b, contained all the parts for one stage, the input RF transformer, T1, the tuning capacitor, C2, the neutralizing capacitor, C1, and the tube and tube sockets. RF coupling to other stages through the filament circuit was prevented by the choke coil, L1, and the capacitor, C3. Similarly, the capacitor, C4, and the resistor, R1, suppressed RF currents from entering the plate supply. Such complete shielding was expensive and was generally reserved for higher priced radios.

The announcement of the Stromberg-Carlson set in Radio News starts by saying:

> Another phase of radio receiver construction which has been getting more attention recently then heretofore is that of

> shielding. . . .without it, it would be impossible to use the three stages of radio-frequency amplification.
>
> (Radio News 1926b)

So it had taken seven years after Hazeltine had drawn his shielded amplifier for shielding to get the attention it deserved.

Dreyer and Manson replaced the old breadboard and bakelite panel by a metal chassis. They provided one knob control by ganging the last three tuning capacitors on a common shaft. However they decided not to gang the antenna tuning capacitor with the others. Thus even with shielding they were unable to get the antenna circuit to track.

Another completely shielded 1926 Neutrodyne was the Music Master Model 250. It used one more RF stage than the Stromberg-Carlson set, the four stages being tuned by one-knob. The five tuning capacitors were all mounted "Grebe style" with their shafts arranged vertically. This arrangement, like Grebe's, allows the designer to use the old chassis layout with the circuits proceeding left to right behind the front panel. Although Grebe used sprockets and chain to gang their three capacitors together, Music Master used levers and rods as described in Chapter 7. The tuning knob was geared to one of the capacitor shafts with a right angled gear so that the tuning shaft projected from the front panel. A small built in loop antenna coil should have made the antenna circuit track but a trimmer capacitor still was provided. Perhaps the public had grown used to this extra control.

Music Master used the Neutrodyne circuit with the neutralizing capacitor connected to a tap on the grid coil, Fig. 4.4b. In addition,

> [the] suppression of oscillations is completed by two small-load circuits coupled to the first and fourth radio frequency transformers. (Laing 1926)

These load circuits reduced the amplification of the stage and broadened the tuning of the RF transformer. The reduction in amplification improved the stability and the broader bandwidth made the tuning less critical and tolerated errors in tracking.

Another unusual set built in limited quantities by Golden-Leutz, Inc., was announced in early 1927. Charles Leutz and Claude Golden are well known for their work in superheterodyne receivers (see Chapter 14) in the early 20's. Having sold his superheterodyne business to Norden-Hauck, Leutz concentrated on the design of TRF sets and by 1927 produced the "Universal Transoceanic Phantom", the "Most Powerful Radio in the World".

Like the Music Master the four TRF stages were fully shielded. But Leutz didn't use Hazeltine's Neutrodyne circuit but used the "losser" method, placing resistors in series with the grids of each stage. It would have been interesting to compare its performance with the Stromberg Carlson set which had only three RF stages properly neutralized to provide optimum amplification. But probably the extra "losser" stage in the Leutz unit more than made up for the inefficiencies of the amplifiers. The four RF stages required five tuning knobs which were too many for a two-handed person to tune all at once. So Leutz provided a method whereby one or all could be ganged and turned together. As in the Perlesz and Ferguson sets described in Chapter 7 the set was constructed in the traditional layout with the stages running from left to right. Instead of mounting the tuning capacitor with their shafts in the traditional front to back way or even vertically as Grebe had done he mounted them horizontally running left to right and connected them with a single shaft. The dials, mounted directly on the shafts had their edges protruding through the front panel, as in the Grebe MU1 (Chapter 7), for thumb operation. But, unlike the Grebe they operated vertically instead of horizontally. The actual coupling between each capacitor and the common shaft was "optional" as the editor of Radio News pointed out:

> The five variable condensers which tune the R. F. circuits are mounted in a line parallel to the front panel. Each is fitted with an individual dial of the vertical type, but they may be grouped together, or "ganged", to simplify the tuning operation. Small couplings between the condensers are simply tightened down with a screwdriver to accomplish this "ganging" in any desired combination. Small "vernier" or midget condensers are connected across the second, fourth and fifth main condensers to allow compensation of these respective instruments when all five condensers are coupled together to turn as one. (Radio News 1927h)

Described as "one of the most spectacular commercial sets in existence" it certainly wasn't for grandma and the baby!

Chapter 11

Audio Frequency Amplifiers

Amplifiers for audio frequency signals were first developed by Bell Telephone for use in their telephone circuits. The first amplifier, called a "repeater", was used in 1913 to make the first long-distance telephone practical. This technology was used in radio receivers to amplify the audio frequency signals from the detector. The amplifiers increased the radio's sensitivity and provided more volume for both headphones and loud speakers. For example, the two tube Radiola III, Fig. 3.5, used a single stage amplifier after the regenerative detector to operate the headphones. All later TRF's used two or three AF stages after the detector to provide sufficient power to operate multiple headphones or a loudspeaker.

The operation of the AF amplifier is based on the principles of the triode amplifier, Fig. 4.1, already discussed in Chapter 4. The amplifier circuit, Fig. 11.1a, is typical of those used in the radios of the 20's. The AF transformer, T1, supplies the input signal to the grid of the amplifier tube. The amplified signal is produced across the primary of transformer, T2, and is coupled to the next stage through the secondary winding. The effect of the grid-plate capacity of the tube is small at audio frequencies and, unlike RF amplifiers, no neutralization is required.

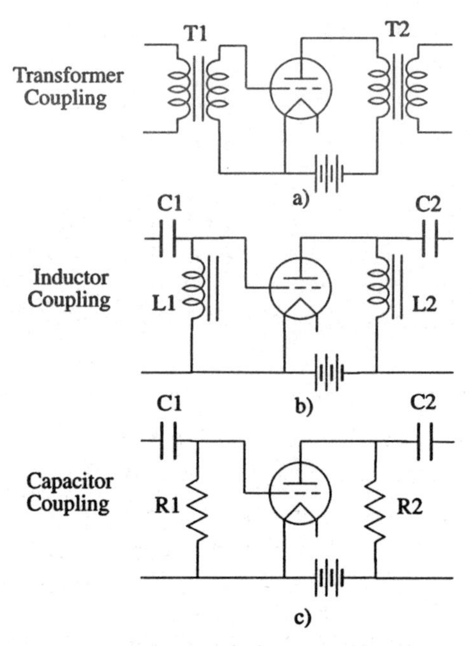

Fig. 11.1 AF Amplifiers

The construction techniques for the AF transformers were developed by the telephone company for their repeaters. They were wound on iron cores to provide sufficient inductance for passing the lowest audio frequency signal (Plate 7). The typical radio used transformers with more turns on the secondary than on the primary. This "turns ratio" stepped up the primary voltage and increased the amplification of the stage. Commonly a 3:1 step-up ratio was used but transformers were later developed with ratios as high as 6:1.

The first AF amplifier followed the detector. The primary of the input transformer was placed in the plate circuit of the detector as shown in the Radiola III schematic, Fig. 3.5. When only one stage was used the headphones were placed directly in the plate circuit in place of the output transformer. The TRF set shown in the schematic, Fig. 15.2, Chapter 15, used a typical two stage AF amplifier with two AF transformers. The last stage operated a loudspeaker placed directly in the plate circuit. By the late 20's the low resistance dynamic speaker was developed and the final stage was provided with a step-down transformer to drive the speaker.

When large signals are amplified, as in later stages of the amplifier, the grid must be provided with additional bias by placing a "C" battery in series with the transformer secondary. The largest signal was amplified in the final stage and a "C" battery was commonly provided for that stage alone. But normally previous stages were biased without a "C" battery by returning the grid circuit to the negative side of the filament, as was the practice for RF amplifiers.

An AF amplifier must be able to amplify all audio frequencies as equally as possible. If the windings resonate at certain frequencies these frequencies will be amplified more than others and the original signal will be distorted. Ideally the response of the amplifier should be the same for all frequencies. Modern stereo amplifiers amplify all audible frequencies equally from 20 to 20,000 Hz providing a nearly ideal response. Such a wide response was not only un-obtainable with the early transformers but was not necessary. Early broadcast stations used carbon microphones similar to the ones in a telephone mouthpiece. These microphones and the early headphones and loudspeakers only responded to frequencies used by the telephone, about 300 to 3000Hz. So transformers developed for telephone use were suitable for the early radios.

However transformer development kept pace with the invention of better microphones and speakers. By 1928 American Bosch was advertising that their radio, using special AF transformers, had a "Fidelity Curve" showing equal amplification of frequencies from 60 to 5000 Hz. This is still an entirely satisfactory response for modern AM radios.

The limitations of the AF transformer can be overcome by using other methods to couple the amplifier stages. The inductance coupled amplifier, Fig. 11.1b, and the resistance coupled amplifier, Fig. 11.1c, were sometimes used in early radios. The output signal is developed across an inductor, L2, or a resistor, R2, and passed on to the next stage through the capacitor, C2. The stages do not have the gain of the transformer coupled stage as the step-up ratio of the transformer is lost. But, when properly designed, they provide a wide frequency response. By 1930 special tubes were designed to be used with the resistance coupled amplifier and provide high gain. Reliable resistors and capacitors became available and gains as much or more than those of the older transformer coupled stage were achieved without the expense of the transformer.

In describing their completely shielded Neutrodyne Dreyer and Manson (1926) discuss the problems of AF amplifiers. They point out that peaks at high audio frequencies could occur due to capacitative coupling between the output stage and the grid-leak detector. They shielded their detector in a metal compartment to avoid this problem. In unshielded sets care had to be taken to place the output tube as far as possible from the detector's grid leak to avoid these frequency peaks and possible oscillations.

The filaments and grid of the old triode tubes were susceptible to mechanical vibrations. If excited by the sound from a loudspeaker the elements might vibrate at their natural frequency which produced an unwanted "microphonic" signal. The most serious offender was usually the detector tube though even the RF amplifier tubes could also be a problem. Tubes were isolated from vibrations from the chassis or breadboard by mounting the tube sockets on springs. Even when these precautions were taken sound vibrations through the air might have been sufficient to disturb the tube. Dreyer and Manson even recommended felt lined compartments for the detector, an expensive fix that was usually avoided whenever possible.

The AF amplifier depended on a good transformer for its performance. They were a relative expensive part and many manufacturers cut their cost which resulted in some unreliable units. The secondary winding was made of thousands of turns of very fine wire only a few thousandths of an inch in diameter. The entire set would immediately cease to operate if a break developed in the fine secondary wire causing the set owner to miss his favorite program. But many manufacturers made reliable units and millions remain in operation today some 60 years later.

Chapter 12

Reflex Circuits

For a short time in 1923 and 1924 there was a great interest in a circuit that was originally invented in 1917 by Marius Latour in France (Goldsman 1923). This "reflex" circuit allowed one tube to simultaneously act as both a radio frequency (RF) and an audio frequency (AF) amplifier and opened up the possibility of providing amplification with an affordable number of tubes.

The circuit of the Crosley Trirdyn, Fig. 12.1, illustrates the operation of the reflex circuit. The first tube performs the function of an RF amplifier and, at the same time, amplifies the AF signal from the detector. The antenna signal is coupled to the grid of the first tube in the usual way through the tuned circuit, L1 and C1. Unlike the ordinary RF amplifier, the grid circuit also contains the secondary of the AF transformer, T1. This transformer has no effect on the RF operation of the tube because the RF signal is effectively connected to ground through C4. Similarly the plate circuit contains the primary of AF transformer, T2, and radio frequencies are bypassed through C5. Thus, as far as the RF signals are concerned, the two tubes formed a receiver with a single stage of RF amplifier followed by a regenerative detector.

The audio frequency output of the detector is coupled to the grid of the first tube by the transformer, T1. The AF signal is

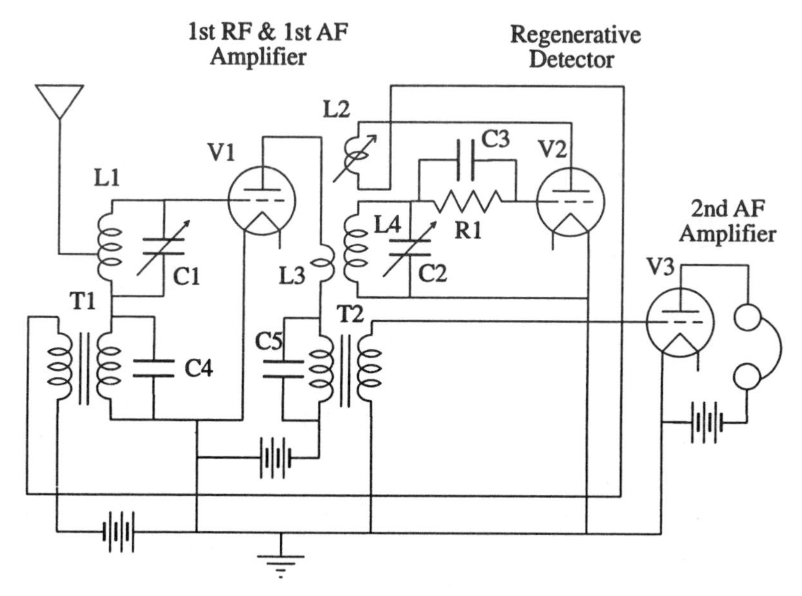

Fig. 12.1 Crosley Trirydn Reflex

not affected by the RF tuned circuit, L1 and C1, or by the relatively small RF bypass capacitor, C4. The amplified AF voltage at the plate of the first tube passes through the transformer, T2, to the final AF stage, V3. Again the inductance of radio frequency coil L3 and the capacitor, C5, had a negligible effect on the AF signal. Thus both the RF and AF signals pass through the first tube and the circuit performs as a four tube set with an RF amplifier, a regenerative detector and a two stage AF amplifier. Thus the Browning-Drake regenaformer, Fig. 4.5, is built with one less tube.

It seems at first sight that a tube has been saved without any change in performance. However, the reflex circuit is not without its own problems. Since both signals pass through the reflexed tube care must be taken that they remain separate and not interfere with each other. If the combined signals become too large the tube may be driven out of its normal operating range. In addition the high resistance of the AF interstage

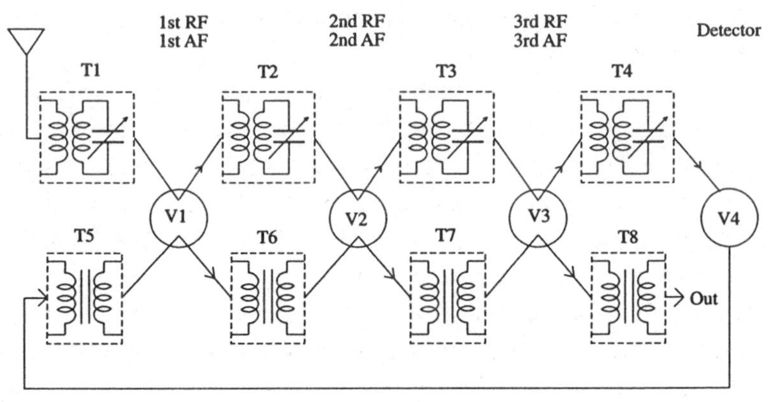

Fig. 12.2 Latour's Reflex

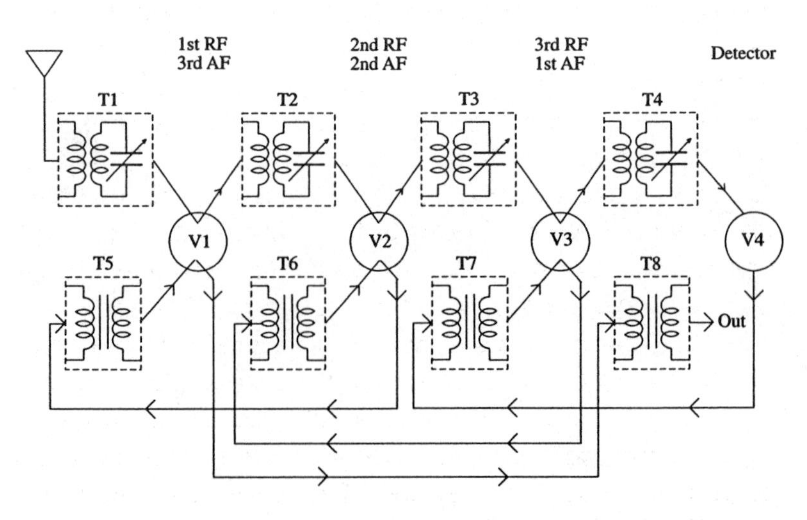

Fig. 12.3 Inverse Duplex

transformers' secondaries may cause them to behave like gridleaks. The reflexed amplifier then no longer amplifies but rectifies the composite RF and AF signal. When this happens, the AF and RF signals heterodyne with each other producing a very distorted output signal.

The above considerations become even more important if many stages were reflexed. Latour extended his concept to the first three stages of a four tube set, Fig. 12.2. The antenna signal was amplified by the first three tubes, V1, V2 and V3, which are coupled by the RF transformers, T1, T2 and T3. The amplified signal was then coupled to the detector, V4, through transformer, T4. The AF signal from the detector was then reflexed back to the first tube and amplified along with the RF signals through the AF transformers, T5, T6 and T7. Thus all three amplifier tubes did double duty amplifying both the RF and AF signals.

The strongest RF signal and the strongest AF signal were both amplified by V3 and it was the first to overload on a strong signal. This and other drawbacks of the Latour circuit were overcome by a reflex circuit developed by David Grimes, a electrical engineering graduate of the University of Minnesota (Durkee 1923). His "Inverse Duplex" circuit, Fig. 12.3, like Latour's circuit, used three amplifier stages, V1, V2 and V3, followed by the detector, V4. However, instead of connecting the AF signal from the detector back to the first tube, V1, Grimes sent it to the third tube, V3. From V3 the AF signal went through T6 back to the second tube, V2, and then through T5 to the first tube, V1. Thus in Grimes' circuit, unlike Latour's, the third tube amplified the weakest AF signal and the strongest RF signal and the first tube amplified the weakest RF signal and strongest AF signal. In this way the signal levels were more evenly distributed between the tubes and they were less prone to overloading.

The Inverse Duplex circuit also overcame another serious problem with Latour's circuit. In Latour's circuit, any residual RF signal passing through to the output of the detector, V4, could be passed back to the first stage through T5 and amplified again by the first three tubes. The four tubes were then acting as a high gain RF amplifier with its output connected to its input producing a large amount of regeneration. Much care had to be taken to insure that the detector output was very small or the regeneration would make the whole set oscillate. Grimes' circuit reduced the possibility of oscillations by connecting the detector output, not to the first tube, but to the tube immediately preceding the detector. Thus the overall gain back to the detector was reduced and oscillations were more easily prevented.

Besides the problems of overloading and RF regeneration Latour's three tube reflex suffered from interference from the household power wiring. This wiring induced a small 60 Hz signal into the antenna. In Latour's circuit the three stages of AF amplification, cascaded from the antenna to the detector, readily amplified these signals adding an interfering hum to the AF signal. The Inverse Duplex cascaded the AF amplifier in the reverse direction so that the first tube, connected to the antenna, was the final AF stage. The 60 Hz power line frequency from the antenna was therefore amplified only by that one stage and the hum was not audible.

A three tube version of Grimes Inverse Duplex was marketed by him in kit form and towards the end of 1923 Sleeper used the circuit in their Monotrol receiver. The receiver circuit used untuned RF transformers between stages relying solely on the tuned loop antenna for selectivity. Thus only one tuning control was required, hence the name "Monotrol". The sensitivity or volume control was provided by a rheostat in series with the tuned loop circuit and the grid of the first tube. This resistance reduced the signal but did not reduce the loop's selectivity. By reducing the strong signals before the first amplifier stage the reflex amplifiers were prevented from overloading.

Other variations on the reflex circuit were generated by an announcement in Radio News of a "$225.00 Reflex Prize Contest" in which they encourage experimenters to build a reflex set around a crystal detector and a single tube. The crystal detector eliminated the need for a separate detector tube and freed the single tube so that it could provide a stage of both RF and AF amplification. Regeneration can even be added to the RF amplifier to increase the sensitivity (Fitch 1923). Many experimenters responded with ingenious designs to try to win the prize. The editors say that they had tested many good circuits but there still must be some that are even better. A commercial set using a crystal detector were made by ERLA in 1923. Their set, the Superflex, used three tubes and a crystal detector (Douglas 1988). The first stage was a straightforward RF amplifier. It was followed by a reflex stage that performed as both an RF and AF amplifier. The third tube provided an additional stage of AF amplification.

The crystal detector in ERLA's Superflex did not provide any amplification and so most manufacturers of reflex sets used a tube detector. De Forest was looking for a company to manufacture radios and in 1922 he bought Radio Craft (Douglas 1988). This company had been founded by Frank M. Squire who had previously been a draftsman at Grebe. Squire was introduced to reflex circuits by William Preiss who had worked on them with the Navy during World War I. Preiss called himself "father of reflex" although Latour already had his patent in 1917. Squire designed the DeForest D7 receiver and it was on the market by late 1922. The set used three tubes reflexed to give the same performance as five tubes. Squire kept improving the reflex sets and brought out new models with more tubes. By 1925 they were making the D17 which was a five tube receiver with one reflexed AF stage (Livingstone 1925). Two RF amplifiers preceded the reflex stage providing a total of three RF amplifier stages. A large loop antenna and tuning capacitor formed the input circuit of the first RF stage which was coupled through a tuned RF transformer to the second stage. The reflex RF stage used untuned RF transformers on the input and output so that only two tuning capacitors were required, one on the loop antenna and one on the output of the first RF amplifier. The reflex stage amplified the AF detector output and the final AF stage drove a horn loudspeaker that was mounted below the set itself. The set, speaker and batteries were all self-contained in an attractive table-top cabinet, a configuration that was to become popular in the 30's.

The reflex circuit had its popularity while tubes and the storage batteries for filament power were expensive but the TRF receivers soon replaced them. They were, however, an ingenious idea and may well have been the forerunner of modern "multiplex" systems that send messages on different frequencies over a single channel.

Chapter 13

Vacuum Tubes From Triodes to Screen Grids

The triode vacuum tube had made all the early radios possible and wasn't replaced until the screen grid tube was developed. Since De Forest invented his triode, the "audion", in late 1906 the triode was the "only game in town". The audion was, by modern vacuum standards, a "gassy" tube because De Forest did not have a good vacuum pump. But fortunately for De Forest the poor vacuum in his audion made it a sensitive detector. In fact, a tube which purposely had a trace of gas, the UV200, was commonly used in detector circuits in the 20's radios.

It was Dr. Harold Arnold of the Bell Telephone Laboratories who made the audion into the modern high-vacuum triode (Tyne 1977 p. 86). As a telephone engineer he was interested in using the audion in a telephone "repeater". The device called a "repeater" originated in the days of telegraphy. Telegraph signals lost their strength when transmitted over long wires and had to be boosted at regular intervals. The boost was given by having a electromechanical relay at the receiving end of the line. The contacts of the relay, connected to a battery, re-transmitted a signal. The relay therefore "repeated" the incoming signal and sent out a new one of original strength. Of course, a mechanical relay couldn't be used retransmit the audio frequencies in a telephone signal so telephonic communication was limited in distance. Many ingenious telephone repeaters were invented but were unsuccessful. A good repeater was needed to make long distance telephony possible.

De Forest demonstrated his audion to Arnold in 1912 and showed off its amplifying capabilities. Arnold saw the possibility of using the audion as a repeater and proceeded to test it for this purpose. He found that the audion was able to amplify but its performance was unreliable. He therefore set about to make it more reliable and he soon realized that the residual gas was the cause of its instability. If the audion was going to be a practical amplifier for a telephone repeater it had to operate with signals that were much larger than those appearing in De Forest's radio detectors. The larger signals required a higher plate voltage than that used by De Forest but when the audion's plate voltage was increased the gas ionized and a blue glow appeared inside the tube. The ionized gas caused the grid to loose control and the audion was unable to perform as an amplifier.

Arnold decided to make a new tube with a high vacuum. This was no simple undertaking as he had to order a high-vacuum pump all the way from Europe as none were available in the Bell Laboratories. By October 1913 he had a working amplifier which he tried out on a local telephone line. This was the first repeater or practical AF amplifier. The tube was called a "telephone repeater element" or a "repeater bulb". This name remained in the telephone community until 1922 when the term "vacuum tube" finally came into common use. When the first New York to San Francisco long distance line was opened in January 15, 1915, the "repeater bulb" made it possible.

The Bell company's subsidiary, Western Electric, made the tubes for the telephone repeaters. Since the telephone company used these tubes themselves and had no other market for them they were made to meet their special requirements. The most important characteristics were a long life and rugged construction. The cost of production was secondary to these goals and the tubes were carefully made by hand. When the United Sates entered the war in 1916 the Navy required rugged tubes for their ship radios. Shipboard use exposed tubes to vibration and shock more severe than that demanded of the repeater bulb. All the manufacturers of tubes geared up to produce the more rugged tubes, including Western Electric, De Forest, and General Electric. The Navy needed tubes in large quantities and placed an initial order of 80,000 tubes with GE and another order for 20,000 tubes in 1918 (Stokes 1982 p. 13). General Electric was in a good position to produce these as they already had a production facility for electric lamp bulbs.

If it had not been for these large war orders the production of tubes during this time might well have been limited and the triode tubes may not have been ready for the broadcast boom in the 20's. Before the war the commercial production of tubes had been stopped by a controversy over patent rights. The inventor of the two-element diode tube, the Englishman John Fleming, worked for the Marconi Company which also did business in America. Marconi held the patent rights to the two element diode while AT&T had previously bought De Forest's patent on the triode. Since the triode is really a diode with an added grid, AT&T couldn't manufacture triodes without infringing on Marconi's patent and Marconi couldn't manufacture them without AT&T's patent. It was a stand off. A court decision in 1916 effectively banned the production of triode tubes except for use in telephone repeaters.

After the war the standoff was resolved when RCA bought out the American Marconi Company and paved the way for the cross-licensing agreement between AT&T, RCA, and GE in July 1920. Westinghouse joined the group a year later forming a monopoly that was only broken up in 1930 by a government anti-trust suit. With this agreement RCA could now distribute tubes made by GE and Westinghouse for the infant radio market. These tubes were built to a common standard and became the ubiquitous triode we have encountered in all the early radios.

The development of new tubes by these large corporations was only carried on if the marketplace demanded change. Small companies were discouraged by the patent situation. The first tubes sold for the radio receiver use, the UV200 and UV201, advertised in 1920, utilized tungsten wire for the filaments. Tungsten is not the best emitter of electrons and the filaments had to be operated as bright as the old lamp bulbs. All the radio sets built at this time incorporated viewing ports in the front panel so that the operator could adjust the filament current to obtain the proper brightness. The tubes were designed with a filament that would operate from 5 volts, the lowest voltage provided by a 6-volt storage battery just before it requires recharging. A series rheostat provided adjustment of the filament voltage as the battery ran down. The tungsten filaments required a current of one ampere at their operating voltage. As we have seen in earlier chapters, when RF and AF amplifiers became popular in 1922 the number of tubes increased from one to as many as four or five. As the tubes increased so did the drain on the set owners batteries and they ran down very quickly. Thus a tube with a lower current filament tube was needed.

The famous scientist, Dr. Irving Langmuir of GE, worked on the emission of filaments throughout 1921. He found that the element thorium was a better emitter than tungsten and he developed the "thoriated tungsten" filament. In early 1923 GE had a UV201A tube in production with a thoriated filament that only drew one quarter of an ampere, a fourfold improvement over the earlier tubes. This tube was used in all storage battery operated sets throughout the 20's.

The only other major tube development was pioneered by Westinghouse for their Aeriola regenerative sets. This tube, the WD11, had a filament designed to operate from a single dry cell and could be used in portable sets. The low current filament was made possible by a further increase in electron emission provided by a coating of a rare-earth oxide. The WD11 was introduced for sale in 1923. At the same time GE introduced the UV199 which had a 3 volt filament designed to operate from three dry cells in series.

Thus the early radio designers had essentially only two types of tubes, the storage battery operated types like the UV201A and the dry-cell types like the WD11 and UV199. The only real difference between them was in the design of the filament so they all gave approximately the same performance. Although some tubes with greater current handling characteristics were produced for driving loudspeakers the radio receiver designers had to wait until the late 20's for a really new tube. This new tube got its start in 1924 when A. W. Hull of General Electric was testing Armstrong's superheterodyne circuit.

Hull was trying to measure the electronic noise generated by the superhet's first detector tube. This noise is caused by random fluctuations in the tube's electron current and is higher when the tube is used as a first detector than when it is used as an amplifier. Hull needed a high gain amplifier so that he could amplify the noise sufficiently to make his measurements and in the course of developing this amplifier he experimented with different tube designs. He added a second grid between the grid and plate of the triode and operated the new grid at a positive fixed potential. This "screen grid" provided shielding between the grid and plate and practically eliminated the need for the neutralizing circuits required by the triode. Not only was neutralization not required but the screen grid tube was capable of much higher amplification. Signals appearing on the plate of the screen grid tube, unlike that of the triode, had little effect on the electron stream. This provided an equivalent resistance, the "plate resistance", very much higher than the triode. The output circuit's efficiency was reduced by the triode's low plate resistance and was therefore able to perform much better with the screen grid tube. In a study of superheterodyne amplifiers (Snow 1924) a triode Neutrodyne circuit was measured to have an amplification of only 10 to 15 times per stage. But Hull was able to obtain an amplification of as much as 60 times per stage with his screen grid tube.

Hull announced his discovery by presenting a paper before the American Physical Society in 1924 (Hull 1924). A short summary of the paper appeared in the Physics Review buried among other papers that had been presented at the meeting. Thus the physics community did not recognize the importance of Hull's invention of the screen grid which really should rank with De Forest's invention of the triode as a milestone in electronics. Fortunately, GE saw its potential but it took them until February 1927 before they could start producing a screen grid tube in volume. Finally, RCA announced a screen grid tube for sale in October 1927. This tube, the UX222, had a 3.3 volt filament which put it into the same class as the UV199 designed for dry cell operated battery radios.

Kits using the UX222 were offered to experimenters in 1927. One of these was made by National, the designers of the coils for the Browning-Drake receiver (Chapter 4). The circuit, Fig. 13.1, was very similar to the Browning-Drake circuit except the first tube has been replaced by the new screen grid tube. Just as with the triode, the input signal from the antenna coil was applied to the first grid and controlled the electron flow from the filament. The screen grid operated at a positive potential somewhat less than the plate potential. The electrons from the filament were attracted to the positively charged screen grid just

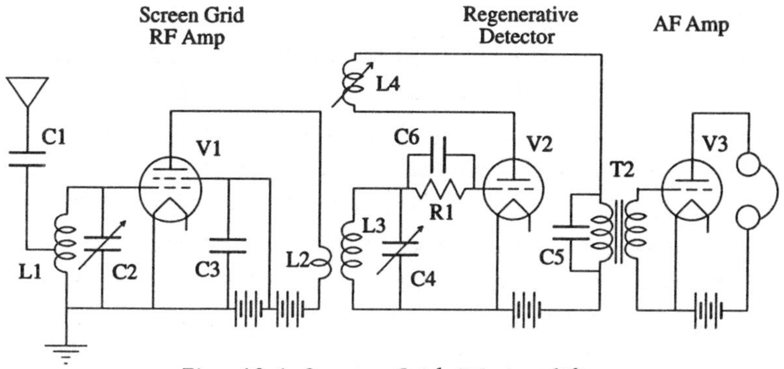

Fig. 13.1 Screen Grid RF Amplifier

as they were attracted to the plate of the triode. But, unlike the plate of the triode the screen grid was kept at a constant potential, the capacitor C3 providing a direct path to ground for the RF signal current. Most of the electrons, approximately two-thirds in the UX222, went through the screen grid to the plate and flowed through the RF transformer as in the triode amplifier. It was no longer necessary to neutralize the stage as the plate was shielded from the grid by the screen grid. The Radiola 21 by RCA was one of the few commercial, battery operated, screen grid sets using the UX222. In 1929 there were still enough homes that were not electrified, mainly on the farm, to justify this set design. Two screen grid RF stages were followed by the usual triode detector and two AF amplifiers. The triodes were a new type, the UX112A, which was developed for higher power output then the old UX201A. They both had 5 volt filaments so an additional series resistor was used to drop the voltage to 3.3 volts for the UX222 filaments. The three tuning capacitors were ganged for one knob control.

One set, a small portable called the "Tom Thumb" made by Automatic Radio, used the UX222 with its 3.3 volt filament companion, the UX199. The latter tube, like the UX222, was designed to operate from three dry cells and had been around since the early 20's. By combining the two types the "Tom

Thumb" was using the tubes in the way their designers intended. The little portable used a loop antenna, one screen grid RF amplifier, a triode detector and two triode AF stages. Unlike the screen grid sets described above it did not need a storage battery and was truly portable. The UX222 must have given this four tube set the sensitivity of a five tube all triode set.

Crosley and Stewart-Warner used the UX222 in sets with three screen grid RF stages that were designed to operate from 110 volt DC power. Its surprising that DC power, which was originally used by Edison in his first electric light systems, was still common enough in 1929 to provide a market for these radios. The Crosley Model 60S used a fourth UX222 for a detector followed by two UX112A's in parallel driving a push-pull power amplifier using two UX171A's. This arrangement, although it used DC filament tubes instead of AC operated heater-cathode types, foreshadowed the typical set of the 1930's.

After the announcement of the UX222 it took two more years for the development of an AC screen grid companion to the UY227, the UY224, which was immediately used in all the AC sets. In the following years engineers, no doubt inspired by the success of the screen grid tube, made a concerted effort to design tubes with even more performance. In a couple of years another grid was added to make the "pentode" tube and then followed the hundreds of tubes developed throughout the 40's, 50's and 60's until the transistor replaced them all. Looking back at the tube's history it seems that a log jam had been broken in the vacuum tube's development releasing the solitary high-vacuum triode from its 15 years of service from 1913 to 1927.

Chapter 14

The Superheterodyne

The year 1917 found Edwin Armstrong, the inventor of the vacuum tube oscillator and regenerative detector, enlisted in the army and working in France on a secret project. The Germans were reported to be using radio frequencies between 500 and 3,000 KHz and the allies wanted to listen in. What they needed was a sensitive receiver that would pick up these frequencies which were, at the time, much higher than those usually encountered in radio communications. If only a vacuum tube amplifier could be made to amplify these waves a sensitive receiver could be constructed. However, as described in Chapter 4, the grid-plate capacity of the triode tubes would cause such high frequency amplifiers to be unstable.

In 1917 amplifiers were in common use in telephone repeaters (Chapter 11) but they only had to amplify frequencies less than 5 KHz. Amplifiers had also been constructed for the long radio waves then generally used for communication up to 100KHz. Above that frequency the amplifiers would oscillate and it would not be until 1923 that Hazeltine found the solution by inventing the Neutrodyne circuit. So Armstrong seemed to be faced with an impossible problem.

Before Armstrong arrived in France, Harry Round, an English engineer working for Marconi, and Marius Latour, the French physicist who invented the reflex circuit (Chapter 12) had been working throughout the war on high frequency amplifiers. As Armstrong says in his paper to the Institute of Radio Engineers:

> Round in England and Latour in France, by some of the most brilliant technical radio work of the war, succeeded in producing radio frequency amplifiers covering the band from 500,000 to 1,000,000 cycles and tho [sic] covering a much more limited band, amplifiers operating on 2,000,000 cycles had been constructed. (Armstrong 1924)

These amplifiers made use of special tubes with very low grid-plate capacitances which were developed in England and France. Armstrong points out that no attention was paid to this problem in the United States as the design of high-frequency receivers for short waves was ignored.

So how was Armstrong going to amplify frequencies as high as 3,000 KHz when the best amplifiers available could amplify only up to 100KHz? The answer now seems simplicity itself. Just change the high frequency to a lower frequency that can be amplified. Fesseden had discovered sometime earlier that if he took two independent oscillators each operating at a different frequency a third frequency could under certain circumstances be generated. He called the new frequency signal the "heterodyne" from the Greek "heteros" meaning different. Fesseden found that the heterodyne was always the sum or difference of the two original signals. A similar effect is used by a piano tuner when he hears the beat note produced by two sound waves—one from his tuning fork and the other from the piano. The autodyne circuit, discussed in Chapter 3, made use of this heterodyne or beat frequency to enable the detection of CW signals. This heterodyning will, in principle, work with any two frequencies to produce a difference frequency.

Armstrong's superheterodyne used the same heterodyne principle but the difference signal he produced was well above the range of audio frequencies. This supersonic signal gave the

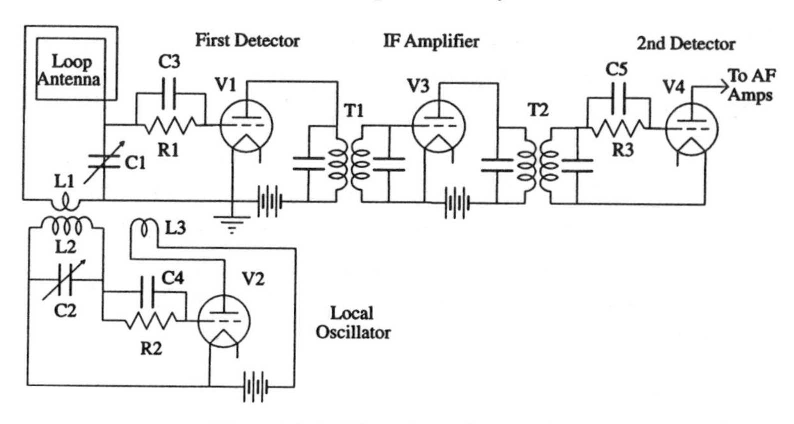

Fig. 14.1 The Superheterodyne

superheterodyne its name. The incoming high frequency RF signal from the antenna was heterodyned with the signal from a "local oscillator". The difference or "intermediate frequency (IF)" so produced was then amplified and detected in the ordinary way. In Armstrong's first set he adjusted the local oscillator signal to produce a 100 KHz heterodyne with the received high frequency wave. For example, in order to receive a 1,000 KHz signal the local oscillator was adjusted to 1,100 KHz and the heterodyne action produced a 100 KHz intermediate frequency. Of course a 900 KHz signal would also heterodyne with the 1000 Khz local oscillator to produce the same 100 KHz intermediate frequency. This unwanted characteristic of superheterodynes has the effect that two radio stations may be received at the same time. The unwanted signal, or "image", was removed by a sharply tuned antenna circuit that selects only one of the incoming signals.

Armstrong's 1918 superheterodyne circuit is shown in the simplified schematic, Fig. 14.1. The loop antenna was tuned by C1 to pick up the incoming RF signal which was connected to the first detector tube. The local oscillator, tuned by C2, generates the local RF signal which was added to the antenna signal through the coupling coil, L1. The local oscillator circuit was essentially Armstrong's regenerative detector (Chapter 3) adjusted to produce continuous oscillations. The first detector used the normal vacuum tube detector circuit but a transformer tuned to the intermediate frequency replaced the headphones. The first detector circuit with its rectifying action was required to produce the heterodyne. If the two signals were connected instead to an ordinary amplifier stage both of the signals would appear at the output unchanged and there would be no heterodyne signal. The amplifier would then be performing the role of a reflex circuit (Chapter 12) amplifying two signals at the same time. The intermediate frequency (IF) was passed through the IF transformer to subsequent amplifying stages.

The AM modulation of the incoming RF signal was not affected by the heterodyne process and the IF frequency signal carried the same modulation. After amplification by one or more IF amplifier stages the IF signal was detected by the second detector and the resulting AF signal amplified as in the normal TRF set. The tuning of the IF amplifier was preset to one frequency at the factory and was not tuned as in the TRF sets. The transformers could therefore be optimized for amplification at the IF frequency. In most superheterodynes of the 20's regeneration invariably occurred in the IF amplifier that helped increase the gain. Some had a "regeneration" control that adjusted the amplification of the tubes by changing their grid bias. Others eliminated the regeneration by using neutralizing circuits.

Armstrong's 1918 "superhet" used eight tubes—the first detector, the local oscillator, three IF amplifiers, a detector and two AF amplifiers. This was a lot more tubes than most people could afford. The two tube set with an RF amplifier and regenerative detector were then popular and later TRF sets only had five or six tubes. By 1922 Armstrong and his skilled assistant Harry Houck made a completely shielded superhet and added an RF amplifier to make a total of nine tubes. The tube count was

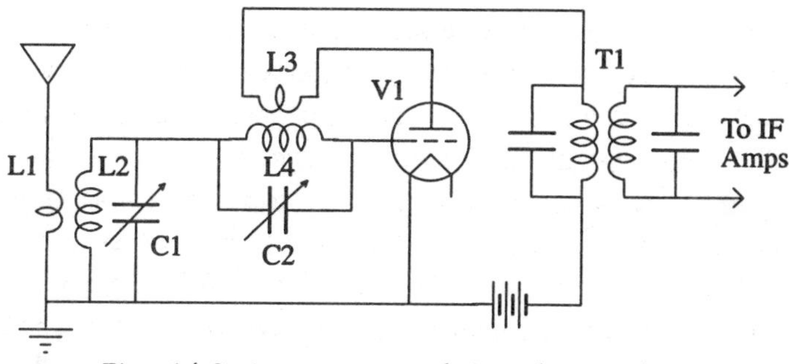

Fig. 14.2 Armstrong and Houck Autodyne

going up, not down, in the wrong direction for an affordable superhet. Before a superhet could be sold to the public the number of tubes had to be reduced and so in 1922 Armstrong and Houck set out to reduce the number of tubes and in less than a year they had a working five tube set.

The first change they made was to combine the first detector and the local oscillator in one tube producing a circuit which was similar to the self-oscillating autodyne detector (Chapter 3). However, they made an ingenious change in the circuits operation which came about as follows. Their self-oscillating circuit, Fig. 14.2, operated by providing regeneration through the tickler coil, L3, to the tuned circuit, L4 and C2. The incoming signal was added to the oscillator signal by the tuned circuit, L2 and C1, in series with the oscillator tuned circuit. In order to produce a low IF frequency the oscillator tuned circuits must be adjusted close to that of the incoming signal. For example, Armstrong's nine tube superhet used an IF frequency as low as 25 KHz, four times lower than the 100 KHz used in his earlier set. In order to obtain a heterodyne of 25 KHz from a 1000 KHz (300 meter) signal the oscillator would have to be tuned to 1025 KHz (1025 1000 = 25). Thus the oscillator frequency differed from the signal frequency by only 2.5 per cent. When two tuned circuits connected in the same circuits are tuned so closely together one will tend to interact on the other. The interaction produced the undesirable effect that caused the tuning of the signal circuit to affect the tuning of the oscillator making it difficult to tune in stations. In addition, the signal from the oscillator, like Armstrong's regenerative detector, was radiated from the antenna. This radiation could cause interference with other receivers.

Houck came up with the solution by tuning the oscillator to half its desired frequency. In the example just given, the oscillator tuned-circuit, L4 and C2, was tuned to 512.5 KHz. Houck then relied on the oscillator to produce a "harmonic" frequency twice the fundamental or 1025 KHz to heterodyne with the incoming signal. Just as a piano string not only vibrates at its fundamental tone or frequency but also has overtones (harmonics) so in an analogous way does the vacuum tube oscillator. Houck had not only reduced the interaction between circuits but had also reduced the oscillator radiation since one-half the signal frequency was greatly attenuated by the input circuit, L2 and C1, before it reached the antenna.

Having reduced the tube count from nine to eight tubes Armstrong called upon his knowledge of the reflex circuit to combine the RF amplifier and the first IF amplifier. Just as other designers had used the reflex circuit to combine the functions of the RF and AF amplifiers Armstrong used the same principles to combine the RF and IF amplifiers in one tube. In the design of a reflex circuit it is important to keep the tube from being overloaded by the combination of the two signals. But, as Armstrong points in his paper, the signals in his reflex circuit are small as the reflex amplifier only has to amplify the relatively small RF and IF signals before they are finally amplified by the later stages of the set. The first two tubes were now providing the function of RF amplifier, local oscillator, first detector and first IF amplifier.

Now Armstrong and Houck's superhet has seven tubes. They then redesigned the IF transformers to produce a higher amplification and with these more efficient units only two IF stages were enough to provide the necessary amplification. Since the first RF amplifier was already doing double duty as an IF amplifier that meant that only one more IF amplifier tube was required. The normal grid-leak detector and a single stage AF amplifier completed their set. Now there were only five tubes, no more than a normal two-stage TRF set.

The RCA, GE and Westinghouse consortium had Armstrong's patents and so the five tube set was demonstrated to David Sarnoff, head of RCA, and he immediately decided that there should be a commercial version. The first commercial superheterodyne radio, the Radiola AR812, was announced in February 1924. An extra AF amplifier was added to Armstrong and Houck's initial design to improve long distance performance so that the AR812 ended up with six tubes. The IF transformers, oscillator coils, AF transformers and tube sockets were all placed a metal box which was then filled with a plastic "potting" compound. This box, referred to as a "catacomb", was not repairable. It is not clear whether this type of construction was done to reduce costs, to hide secret design features or to prevent the serviceman from attempting to fix unfamiliar circuits. Whatever the reason there was little for an interested circuit designer to see behind the front panel except the tuning capacitors and batteries.

The simplified schematic of the AR812, Fig. 14.3, shows how the Armstrong-Houck superhet design was implemented. The first tube, V1, was reflexed to provide both RF and IF amplification and the second tube, V2, performed the dual purpose of first detector and local oscillator. The other four tubes had normal single functions of an IF amplifier, V3, a detector, V4, and AF amplifiers, V5 and V6. L4 and L1 comprised the first IF transformer. The primary, L4, was tuned to the IF frequency by the capacitor, C4, while the secondary, L1, was a high inductance untuned coil. The RF signal from the loop,

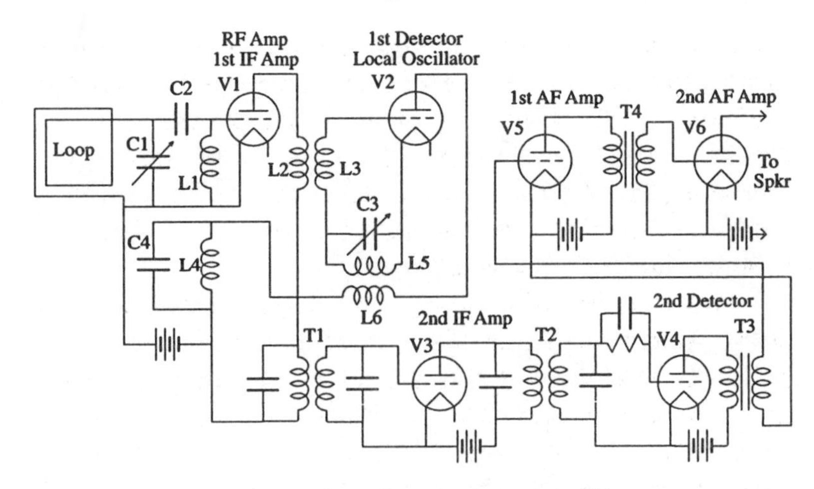

Fig. 14.3 AR812 Superheterodyne

which was tuned by C1, was connected to the grid of V1 through the capacitor, C2. The inductance of L1 was sufficiently high so as to not interfere with the RF signal. The exact value of the inductance was not critical as the RF frequency was more than ten times higher than the IF frequency. The amplified signal was coupled from the output of V1 to the detector-oscillator through an untuned RF transformer, L2 and L3. The local oscillator circuit followed the basic design of the autodyne circuit, Fig. 14.2, the frequency being set by the tuned circuit, L5 and C3, at half the heterodyne frequency. The tickler coil, L6, provided the regeneration required to produce oscillations. The first harmonic of the oscillator generated across L5 was added to the RF signal across L3 and the heterodyne action of V2 produced the IF frequency across the IF transformer primary, L4. This signal was reflexed back through L1 to the grid of V1. The IF signal was then passed through the tuned IF transformer, T1, to the second IF amplifier, V3, and thence through transformer, T2, to the second detector, V4. The AF output of the detector was amplified for operation of the loudspeaker by the two stage AF amplifier consisting of the two AF transformers, T3 and T4, and tubes V1 and V2. A volume control was provided by a rheostat (not shown) that adjusted the filament current of the second IF amplifier, V3. The AR812 was mounted in a portable box with the loop pivoted on top so that it may be rotated for the best reception. The directional characteristic of the loop and its portability made the set popular for direction finding and locating "bootleg" transmitters.

In 1921, while Armstrong and Houck were developing their superheterodyne another engineer, Charles Leutz, who had worked for American Marconi during the war designed and marketed construction kits. His designs were based on Armstrong's initial set design with a separate local oscillator. His Model C used eight tubes to which a two stage regenerative RF amplifier could be added. In his booklet on the construction of his superhets he describes a ten tube set, the Model L, which was

> designed principally for the experienced operator and the proper control of its component parts calls for considerable skill to obtain maximum results. (Leutz 1924)

An idea of the set's complexity can be obtained by considering that there were as many as 22 controls on the two front panels, each panel measuring 40 inches wide and eight inches high. Placing the two units side by side the set would extend to almost seven feet, probably setting a record. Tuning was accomplished using six large dials and four multi-position switches. Leutz took four pages in his book to describe the operation of these controls. On the left of the first unit were three switches which changed the wavelength range of the receiver. The first large tuning knob controlled the local oscillator tuning capacitor. The second knob, again in conjunction with two multipole switches tuned the antenna circuit. The third knob operated a variocoupler to adjust the coupling between the antenna and the RF input coil. The fourth knob tuned the antenna coil while the fifth and sixth knobs tuned the primary and secondary of the first IF transformer. The following IF transformers, five in all, were untuned, thankfully requiring no controls, and were followed by the detector and two stages of AF amplification. Every IF tube had separate filament rheostats which, Leutz says, were necessary to adjust for the great difference in the characteristics of tubes that are "supposedly identical". To top it all off there was the "stabilizer" that adjusted the grid bias of all the IF tubes. When the stabilizer was adjusted to the negative side the amplification was increased but the operator had to be careful for Leutz said that

> the stabilizer must not be moved too far or oscillations will be generated in the amplifier with consequent distortion and loss of signal strength.

Armstrong gave the superheterodyne high marks for requiring only two tuning controls but Leutz found a need for many more. His set was really for "experts only"!

When RCA came out with their superheterodyne in 1924 they guarded their patent carefully against infringement and didn't allow any other manufacturer to use their circuit until the government broke up the trust in 1930. But other companies did get away with selling superheterodyne kits to experimenters just as Leutz had done. During 1924 and 1925 the most famous kits were made by Scott and McMurdo Silver (Silver Marshall). The performance of their sets became legendary when Scott took his ten-tube "World's Record Super" to New Zealand and was able to pick up broadcast stations as far away as Chicago, 8000 miles distant. McMurdo Silver was brash enough to market complete sets but was stopped when a court injunction obtained by Westinghouse was served on him in 1926 (Douglas 1988 v.3 p.83). However he developed a new type of autodyne circuit (Silver 1925b and 1925c) which used a bridge circuit to reduce local oscillator radiation. Silver's "Super-Autodyne" used the principles of the balanced bridge circuit used by Walbert to neutralize RF amplifiers (Chapter 9). The bridge, Fig. 14.4, consisted of two coils, L2 and L3, and two capacitors, C2 and C3,

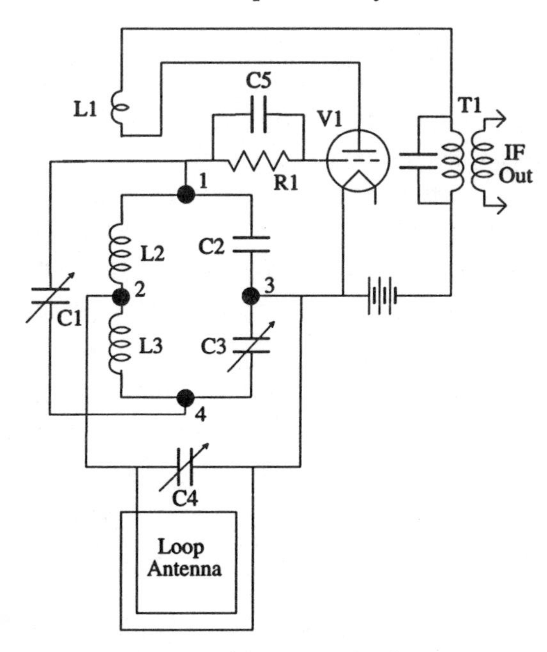

Fig. 14.4 Super Autodyne

which were placed in the input circuit of the oscillating first detector, V1. Regeneration was provided by the tickler coil, L1, coupled to coil, L2. The oscillator frequency was determined by the tuned circuit consisting of the series combination of L2 and L3 in parallel with variable capacitor, C1, and the trimmer capacitors, C2 and C3. So far, except for the center tap on the coil formed by L2 and L3, and capacitors, C2 and C3, the circuit was identical to the usual regenerative detector. But, now, instead of connecting the loop antenna and its tuning capacitor, C4, in series with the grid circuit it was connected across the other two terminals of the bridge, points 2 and 3. When the bridge was balanced by the proper adjustment of C3 the oscillator voltage across points 1 and 4 did not appear across points 2 and 3 to which the loop antenna was connected. Thus there was no signal passed to the loop and consequently the local oscillator signal couldn't radiate from the antenna.

McMurdo Silver's six tube kit was constructed using a bakelite panel about 4 inches high and 17 inches long, smaller than usual for TRF sets but quite adequate to accommodate the two tuning dials of a superhet. A picture in his article shows the set sitting on the front seat of what looks like a Model T. The batteries are set in the back seat and the portable horn loudspeaker is shown on the running board. The picture's caption reads

> The receiver in an automobile. The A battery comes from the automobile using the Lynch Lead. The rather dilapidated bag in the rear holds the B and audio amplifier C Batteries. The Amplion loud speaker and the folding loop also go in this bag when not in use. Blanket-roll straps provide a convenient means for carrying the set itself. (Silver 1925b)

Obviously the set was not meant to be used when the car was going along but only at picnics or at the top of hills were long distance reception was good.

Even though most of these kits used low IF frequencies (40 to 50 KHz) the IF amplifiers still were prone to oscillation. The gain was usually adjusted below the point of oscillation by either reducing the IF tube filament voltage or changing the grid bias or both. The resulting regeneration had the benefit of increasing the amplification and, although it made these kit sets difficult to adjust, the experimenters were able to get the great sensitivity they were looking for. So none of them tried to stabilize the amplifier using the Neutrodyne method except for the amateur McLaughlin who had built the "Super Calamityplex" (Chapter 8). His "Neutrodyne Superheterodyne" (QST 1924) had four shielded Neutrodyne IF stages. The four stages were separated by shielding which as we have seen in our discussion of shielded TRF's was necessary to prevent stray interstage coupling that would upset the neutralization.

By the end of the 20's the screen grid tube and its successor, the pentode, allowed RF and IF amplifiers to be easily constructed. The better amplifiers allowed the IF frequency to be raised to near 450 KHz which reduced interaction between the antenna and local oscillator circuits. Further inventions allowed the oscillator tuning capacitor to be ganged to the antenna capacitor making the one-knob superhet radios of the 30's possible. Today the superheterodyne principle is used in every radio and television set, in the reception of satellite communications and in many other electronic devices. Our everyday use of his circuit is a fitting memorial to the imagination of Edwin Armstrong.

Chapter 15
AC Replaces Batteries

By 1927 most homes in the United States were electrified and there was a great demand for radios that could operate from alternating current (AC) power (Stokes 1982). AC operated power supplies, the "battery eliminators", were already on the market. Those for the filament "A" supply were no more than battery chargers and had to be used with a storage battery. The customer demanded a batteryless set that had a self-contained power supply and could be plugged directly into his power plug. One of the major technical problems that prevented such a set from being economically produced was the high direct current demanded by the battery tube's filaments. Power rectifiers for producing high quality DC were too large and expensive. If, on the other hand, the filaments were run directly from an AC transformer the filament's polarity was switched back and forth at the power frequency. Thus the grid bias was changed at a 60 Hz rate and a loud hum was heard in the speaker.

A receiver that used an ingenious method to provide DC filament power was manufactured by Dictograph in 1926 (Radio News 1926d). The six tube set used standard 5 volt battery tubes to provide two RF amplifier stages, a detector and three AF stages. The filament of each tube, a standard UV201A, drew one fourth of an ampere, a total of one and a half amperes for all six. In order to avoid the costs of such a high current supply the

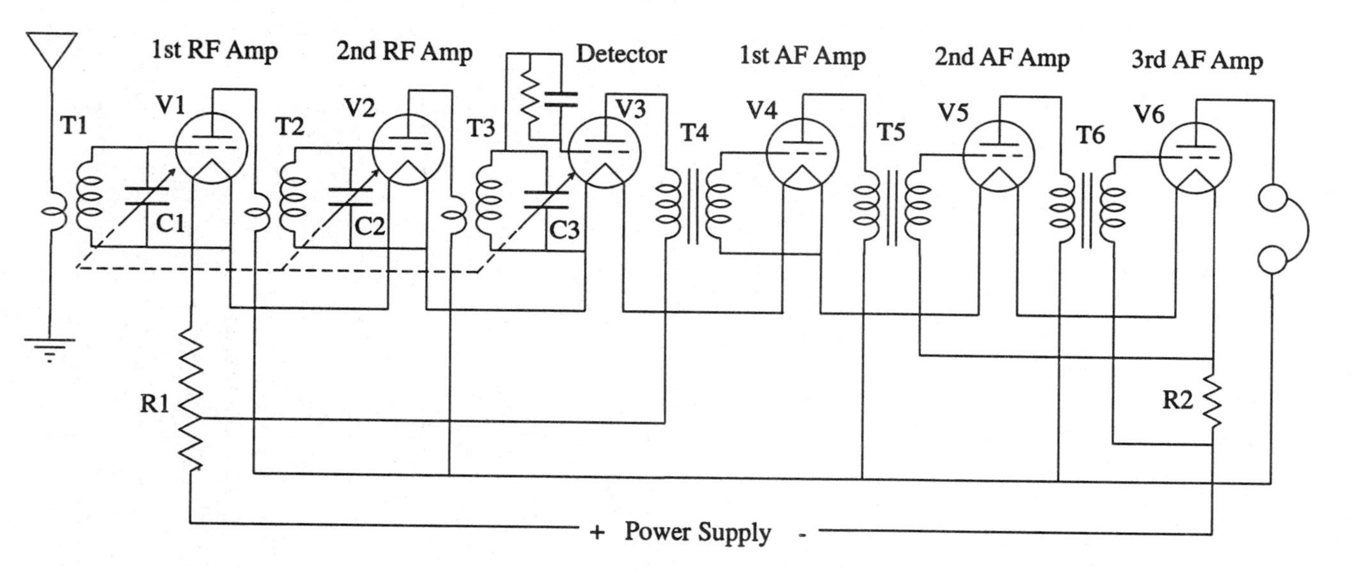

Fig. 15.1 Dictograph

engineers at Dictograph placed all the filaments in series. Thus they needed a 30 volt supply that provided only one quarter ampere for all the filaments. They realized that, by adding a resistor in series with the string of filaments they could operated the entire set from a 45 volt "B" supply. This supply would not only supply the filament current but also provide plate voltage for the tubes and could be obtained from a relatively inexpensive power supply.

The schematic of the Dictograph set, Fig. 15.1, shows the filaments and the resistors, R1 and R2, connected in series across the 45 volt power supply. A tap on R1 supplied a lower plate supply voltage to the detector, V3. The voltage across R2 supplied bias voltage for the final AF amplifier. At first glance it would seem that the bias on each tube would differ by the filament voltage, 5 volts, from each succeeding stage, a total of 30 volts from beginning to end. This much bias would, of course, had made the amplifiers inoperative. But closer examination of the circuit reveals that the designers arrived at an ingenious solution by using transformer coupling for every stage. The secondary or grid coil of every transformer was returned to the proper point in the circuit to provide the correct bias. The RF and the first AF stages had their grid returns connected to the negative side of their respective filaments to supply an average two and a half volt bias, the detector grid was returned to the positive side of its filament to provide a positive bias, the second AF amplifier grid was returned to the positive side of the third AF amplifier's filament and the third AF amplifier grid was returned to the negative end of R2. Every stage was thereby properly biased even though 30 volts existed between the filament of the first and last tube.

Although Dictograph seemed to have had a good method for providing a DC filament current from an AC operated power supply, the connection of all the filaments in series restricted the design of more complex circuits. A more practical design had to operate all the filaments from a step down transformer. It was found that the hum produced by operating the filament from AC could be reduced by using a center tap on the transformer winding. The grid circuit could then be returned to the center tap and the effect of the alternating filament voltage would be cancelled out. This cancellation requires that the filament be symmetrical and worked better when the filament voltage was small. The standard battery tube, the UV201, with its five volt filament did not provide good cancellation of the hum. The voltage across the filament produced varying electric and magnetic fields that affected the emitted electrons even though the applied AC voltage was balanced. These effects were reduced by designing a tube with a low voltage filament. In 1927 the UX226 was announced by RCA for use in AC operated sets. The UX226 filament operated from one and a half volts with a little more than one ampere. The filament power was therefore close to the power drawn by the UV201A and it had similar operating characteristics. However the filament voltage was less than one third that of the UV201A and, everything else being equal, would reduce the hum problem by that factor. The tube was used in many AC sets during the period of transition from DC to AC.

The final solution to satisfactory AC operation lay in the design of tubes wherein the heating function of the filament was separated from its emission function. In this new design, the filament was replaced by two elements. A tightly wound insulated wire coil, the "heater", provided the heat. The heater was surrounded by a cylindrical metal "cathode" coated with electron emitting oxide. Like the filament, the cathode emitted electrons when it was heated, but now the heat came from the heater. The cathode was insulated from the heater so that the latter could be heated by an AC current without having any effect on the cathode. The first heater-cathode tube, the triode UV227, was announced by RCA in May 1927. Now engineers

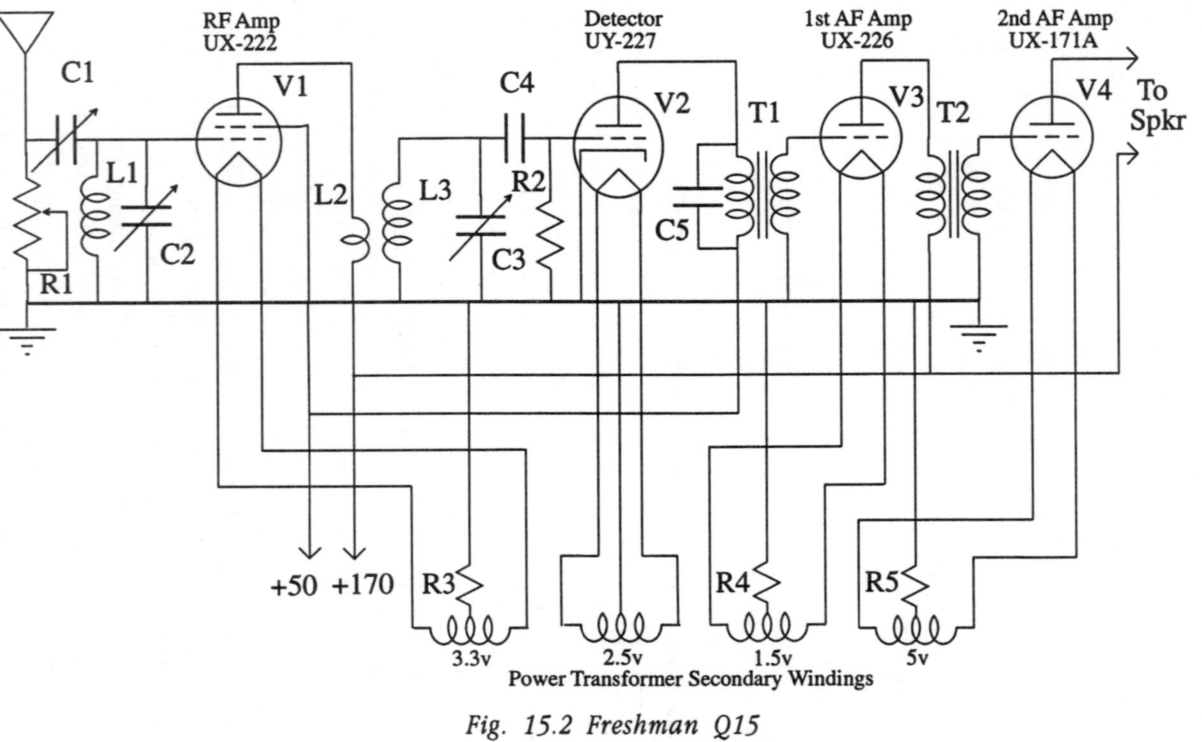

Fig. 15.2 Freshman Q15

had a tube that could be used in AC sets and operate in the same circuits that they had developed for the battery sets. The demand was there and the manufacturers went after it producing AC operated radios with the old triode circuits but with the new heater-cathode tube.

In 1928 Freshman was offering their AC operated Q15 using the new UX222 filamentary screen grid tube, Fig. 15.2. The single screen grid stage was followed by the new UY227 triode detector with a 2.5 volt heater, a UX226 triode AF amplifier with a 1.5 volt filament and a UX171 power output triode with a 5 volt filament. This works out to be a total of three filamentary tubes and one cathode-heater tube each having a different filament or heater voltage. All the tubes could be operated from an AC transformer but each one had to have its own filament winding on the power transformer. Each winding was center-tapped to provide hum cancellation. Resistors R3, R4 and R5 were placed in between the center taps and ground to provide bias for the tubes. Except for the filament circuits the design followed that of the screen grid amplifier, Fig. 13.1. The rheostat, R1, from the antenna to ground served as a volume control. This certainly was a transitional design lying between the older battery radios and the new AC operated radios which only required a single supply for all the tubes heaters.

It didn't take long for RCA to build a superheterodyne using their new cathode tubes. Their Radiola 62 had all the features that later radios incorporated such as one dial tuning and a large console cabinet containing a dynamic speaker. Earlier sets had two dials mounted close together for thumb wheel tuning. But the antenna and local oscillators did not track. The Radiola 62 engineers solved this problem by placing a fixed capacitor in series with the local oscillator tuning capacitor. This compensating "tracking" capacitor made the one-knob superhet practical and has been used ever since. The superhet had finally shed its two controls and had become a true one-knob radio.

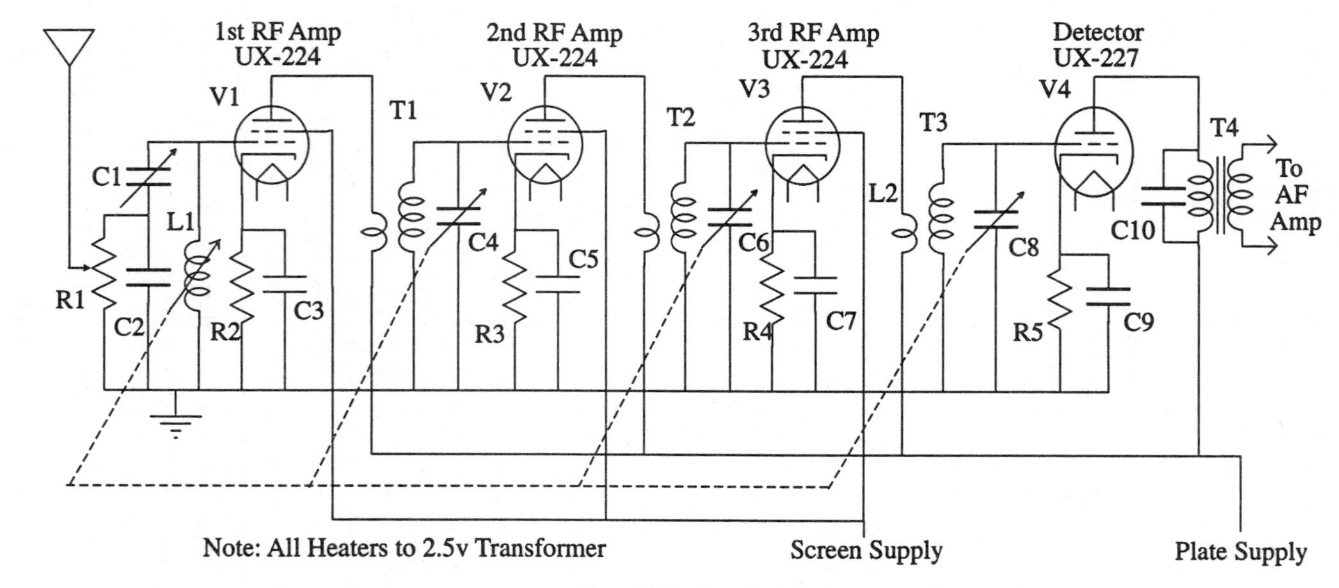

Fig. 15.3 Bosch 48

The Radiola 62 used seven UX227 triodes for the two RF amplifiers, the local oscillator, the two IF stages and the first detector. The designers had to fall back on the filamentary power tube, the UX171A, to drive the loudspeaker as a heater-cathode power tube was not yet available. The IF amplifiers were neutralized with Hazeltines Neutrodyne circuit like McLaughlin's superhet described in the last chapter.

The final development of the TRF radio is exemplified by the screen-grid Bosch Model 48 manufactured in 1929 (French 1929). This unique set was one of the few AC operated TRF's built before the superheterodyne radios of the 30's and deserves to be described here before this story of the 20's radios ends. The design bridged the gap between the battery radios and the new AC radios. A three stage RF amplifier, Fig. 15.3, used the new type UX224 screen grid tubes with heater-cathodes. In order to equalize the amplification over the entire tuning band the antenna was coupled by a tuned variometer, L1. The sensitivity of the variometer tuned circuit was greater at low frequencies when the inductance was higher and compensated for the drop off in response in the following three capacitor tuned circuits. The variometer was a reminder of the crystal and regenerative radios but in the Bosch set it wasn't a separate control but was driven through gears from the ganged tuning capacitors. By using the proper gear ratio the tuning of the antenna circuit closely followed that of the ganged capacitor tuned stages.

Another interesting feature was the detector circuit. The screen grid tubes provided so much amplification that a sensitive grid-leak detector was not necessary and would have been overloaded by the large signals. The detector tube rectified the RF signal in the plate circuit rather than in the grid. This provided a greater signal handling capability and makes an intermediate AF stage unnecessary and a single AF amplifier provided sufficient signal for the loudspeaker. Power amplifier tubes with heater-cathodes were not yet available when the Bosch set was designed so, like the RCA superheterodyne, filamentary types are used instead.

The Bosch set was built on a steel chassis with shields over the tubes, capacitors and coils. The tuning capacitors and the variometer were driven by a worm gear from one tuning knob of the front panel. The antenna trimmer, individual tuning knobs and the old breadboard had been discontinued and would remain so in all future radios. The TRF had reached its final development and was soon to be replaced entirely by the superheterodyne.

Chapter 16

A Glimpse of the Future

The long reign of the triode tube came to an end in 1929 when it was supplanted by newly invented tubes. Most of the circuits described in this book disappeared along with the triode. Gone were the Latour's reflex circuit and Grimes' Inverse Duplex. Also left behind were the Autodyne, the Technidyne, and a host of other "dynes". DeForest's audion, the first grid leak detector, which had served well for twenty years was about to be replaced with a new version of Fleming's 1904 diode. Armstrong's regenerative detector also dropped out of use as regeneration became renamed "positive feedback" and was generally to be avoided. Hazeltine's Neutrodyne that made triode RF amplifiers practical was made superfluous by the screen grid tube. The proliferation of other circuits designed to stabilize the ubiquitous triode, the lossers, the Equaphase, the Counterphase and even Tuska's Superdyne were now part of history. The hard work that was put into the one-knob receivers was forgotten as true one knob control was established for all radios. Out of all the designs with belts, rack and pinions, chains, worm gears and clutches Chamberlain's one shaft won out. The ganged capacitor was now standardized as part of every radio. Hazeltines magic angle of 57 degrees for the axes of the TRF coils was now a mute point as all sets were shielded. The coils were placed in cans as Walbert had done in his 1926 set. Hazeltine's 1919 proposal for individual shields around each stage came to pass. The listener no longer had to put up with the expense of numerous batteries as AC operation from household current was the rule, not the exception. New loud speaker designs made the design of audio frequency amplifiers more critical. The old workhorse, the 3:1 audio frequency transformer, was dropped in favor of resistance-capacitance coupling. New tubes made it easy to provide all the amplification a broadcast receiver designer would ever require.

But one important "dyne" still remained and that was the superheterodyne. It had taken second place to the TRF's during the 20's in spite of years of work by Armstrong and Houck. But it came into its own in the 30's. Its success had to wait for the new tubes, the pentodes, hexodes, pentagrids and a host of others that simplified RF amplification and allowed the local oscillator and first detector to be efficiently combined in one tube. While Armstrong's goal was to provide a low frequency IF signal that could be amplified with the triode, the new RF amplifiers could amplify radio frequencies without any trouble. Now, the primary purpose of the IF amplifier was produce the selectivity required to separate the growing number of radio stations. The intermediate frequency was raised over ten times, from Armstrong's 45 KHz to the new standard 455 KHz. The image signals were negligible and stations no longer came in at two places on the dial. Four tubes plus a rectifier for the power supply supplied greater performance than a six tube triode TRF. A single tube performed the oscillator and first detector function while also providing amplification. A single IF stage provided the gain of two early stages. A special diode-triode provided detection and sufficient AF amplification to drive a power pentode tube.

The electronic age followed the age of the triode, beginning in the 30's and spurred on by World War II. The old triode circuits were the fore-runners of radar, television and computers. We owe all this in great part to the designers of the old 1920 radios. All of them, without the advanced technical knowledge we now have, built successful radios that made it possible for millions of set owners to listen. They brought instant music and news to everyone world-wide. That, in itself, was truly a great technological revolution.

Appendix

Symbols and Units

The electronic symbols used in the schematics are defined in Fig. A1. Most schematics are simplified by omitting the circuits used to supply current for heating tube filaments. They also repeat the plate supply "B" battery for each tube, when in a practical set they would be combined in a single battery.

The names of electrical units and components have changed since the 20's and one has to keep this in mind when reading the old texts quoted in this book. The following table compares the old and the new:

UNIT	NEW	OLD	NEW	OLD
Farad	Capacitance	Capacity	Capacitor	Condenser
Henry	Inductance	Inductance	Inductor	Coil
Ohm	Resistance	Resistance	Resistor	Resistance
Frequency	Hertz	Cycles		

Note the abbreviation, cycles, for cycles per second or Hertz.

Usually the old texts used wavelength rather than frequency when discussing radio frequency waves. The formulas for converting from meters to frequency are:

Meters x Hertz = speed of light = 300,000,000 meters/second or MHz (1 million Hertz) = 300 ÷ Meters and Meters = 300 ÷ Mhz

Handy examples: (KHz = 1000 Hertz)

20 Khz	50,000 meters
100 Khz	3,000 meters
500 Khz	600 meters
1000 Khz (1 Mhz)	300 meters
1500 Khz (1.5 Mhz)	200 meters
5000 Khz (5 Mhz)	60 meters

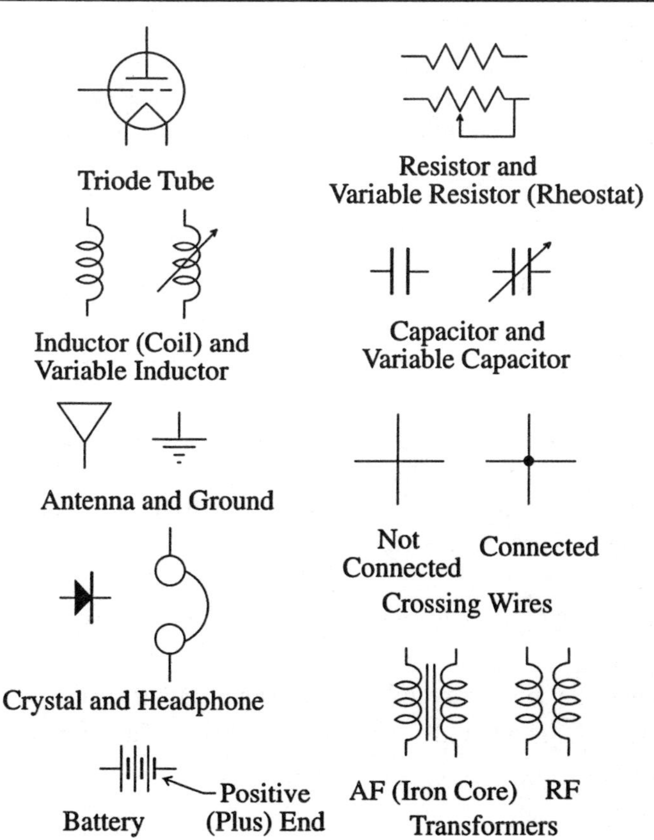

Glossary

Technical Terms

AC. See Alternating Current.

AF. See Audio-frequency

Alternating Current. An electric current that alternates from a positive to a negative direction usually at a definite frequency. The abbreviation, AC, is also used to designate an alternating voltage as in AC Voltage.

Alternator. A rotating machine for the generation of AC currents.

AM. See Amplitude Modulation.

Ammeter. A meter that indicates the flow of current (amperes) in an electric circuit.

Amplifier. A vacuum tube device for increasing the amplitude of an electrical signal.

Amplitude Modulation. The modulation of the amplitude of a radio signal. See Modulation.

Arc. A sustained electric discharge between two conductors.

Audio-frequency. Frequencies lying in the audible range approx. 20 to 20,000 Hertz.

Bakelite. Originally a trade name. A black plastic used in 1920 radios as an insulator. It can be formed into sheets and moldings.

Balance. A circuit consisting of two paths is said to be in balance when currents or voltages are equal in each path.

Battery. A series of electrical cells for supplying electrical voltage and current. In the old radios, an "A" battery supplied the filament current, the "B" baterry the plate voltage and the "C" battery the grid bias. The "A" battery was usually was made of rechargable lead-acid cells while the "B" and "C" batteries were made from a large number of dry cells.

Bias. A steady voltage applied to the grid of a tube in series with the signal.

Breadboard. A wooden base board used by early experimenters and radio manufacturers to mount the radio components.

Bridge. An electrical circuit consisting of four arms each containing a resistor, inductor or capacitor. The input and output voltages across opposite sides have no effect on each other when the bridge is balanced.

Capacitor. An electrical component exhibiting capacitance. It usually consists of two paralled conducting metal plates separated by air or an insulator.

Cathode. The negative electrode of a vacuum tube which emits electrons when heated by a "heater".

Coil. A winding of electrical wire usually in the form of a helix that makes up an inductor or a transformer.

Component. An individual electric unit such as a vacuum tube, capacitor, inductor, resistor, etc. Also any mechanical unit such as a knob, gear, panel, etc.

Condenser. Old name for a capacitor.

Control. The device, usually a knob, for controlling a function or electrical current such as a tuning control, filament control, volume control, etc.

Coupler. A device for coupling an RF signal from one part of an electrical circuit to another. Specifically the arrangement of two coils placed so that the RF electric current in one is electromagnetically transferred or coupled to the other.

Coupling. The electromagnetic interaction between two parts of a circuit whether it be made purposely (see Coupler) or is inherent in the proximity of the components (Stray Coupling). Coupling of RF signals may be due to capacitance, inductance, resistance or all three combined.

Crystal. A sensitive crystalline mineral that, when provided with a sharply pointed wire pressing lightly against its surface, rectifies or detects RF currents.

Detector. A device for indicating the prescence of radio waves or RF currents. A demodulator to recover the audio signal from an amplitude-modulated RF signal.

Diode. A two element tube having a filament and a plate. A diode exhibits a rectifying action.

Direct Current (DC). A steady electric current as opposed to alternating current. A steady electric voltage, "DC Voltage". The output of a battery.

Feedback. The transfer of a signal from the output of an amplifier to the input either through a circuit component or through stray coupling..

Filament. The negative electrode of a tube, usually a thin wire, which is heated by passing current through it causing it to emit electrons.

First Detector. A detector used in a superheterodyne to provide the heterodyne action between the received signal and the local oscillator.

Frequency. The rate that an AC signal alternates measured in cycles per second or Hertz. In old radio jargon a "cycle per second was shortened to "cycle" as in "kilocycle".

Gain. The overal amplification of an amplifier. The ratio of output to input signal voltages.

Grid. The second intermediate electrode of a triode that controls the electron flow.

Harmonic. The frequency component of a complex AC signal that is an integer multiple of the fundamental frequency.

Heater. In a tube with a cathode, the wire coil that heats the cathode.

Heterodyne. The difference frequency produced when two signals of a different frequency are combined in a detector.

IF. see Intermediate Frequency.

Induction. The production of a current or voltage in another circuit due to electromagnetic coupling particularly between two inductors or coils.

Inductor. An electrical component exhibiting inductance. Usually a coil of wire wound on an insulated form (air core) or an a laminated or powdered iron core.

Intermediate Frequency. In a superheterodyne receiver the difference frequency provided by the heterodyne between the input signal and the local oscillator. Intermediate in frequency between RF and AF.

Local Oscillator. The oscillator in a superheterodyne receiver that generates the RF signal that heterodynes with the received RF signal.

Loop. An antenna made of a large coil wound on either a circular, square or triangular forms usually more than one foot across. An inductor made of such a coil.

Loss. The dimunition of amplification, the opposite of gain. "Losses" are electrical parameters, usually resistance, placed in a circuit, either purposely or accidentally, that impair the performance of the circuit.

Modulation. The changes in an RF signal that carries the information to be transmitted. see Amplitude Modulation.

Multistage Amplifier. An amplifier using many single tube amplifiers in cascade to produce a higher overall gain.

Neutralize. Remove or nullify the effects of the grid-plate capacitance of a triode vacuum tube.

Oscillations. Internally generated RF or AF currents in an electronic circuit or amplifier either purposively caused by the design of the circuit or do to indadvertent electromagnetic couplings between components.

Panel. The upright bakelite or metal plate upon which the radio controls are mounted. Usually attached to the breadboard.

Pentode. A five element vacuum tube having three grids.

Plate. The anode or positive electrode of a vacuum tube.

Power Supply. A supply of DC voltages obtained from the AC consumer power replacing the batteries of earlier radios.

Primary. In a transformer the winding to which the input signal is applied.

Radio. A receiver for receiving broadcast transmissions. General communication by electromagnetic radiation or "radio waves".

Radiofrequency. In the early days any frequency above 15 KHz that could be radiated from an antenna. Today limited to frequencies above about 100KHz. Also refers to the signal received by broadcast radios.

Receiver. Any electrical circuit containing crystals, tubes, etc. for the reception of radio wave transmissions. A Radio.

Rectifier. A device that passes electrical current in one direction only. A device to produce DC current from AC current.

Reflex. An amplifier for amplifying signal of two different frequencies at the same time. A signal is said to be "reflexed" when it is returned to a previous stage to be amplified along with another signal.

Regeneration. The increased amplification produced by augmenting the input signal of an amplifier by part of the output signal. Positive Feedback.

Resistor. An electrical component that exhibits resistance, usually made of high resistance wire or carbon.

Resonance. The state in which a tuned circuit responds maximally to an AC signal.

Resonant Frequency. The frequency at which resonace occurs.

RF. see Radiofrequency.

Schematic. A line drawing using electrical symbols showing the connections of the components that make up an electrical circuit.

Screen Grid. The second grid in a screen-grid tube that shields or screens the plate from the first grid.

Second Detector. The detector in a superheterodyne that demodulates the signal.

Secondary. The winding of a transformer that provides the output signal that was induced by the primary winding.

Selectivity. The sharpness with which a tuned circuit or receiver will tune to an incoming signal. The ability to seperate radio stations transmitting on closeby frequencies.

Sensitivity. The ability of a receiver to receive weak signals. A measure of the useful amplification of a receiver.

Set. A complete radio. A set of parts to make a radio.

Shielding. Metal plate or screen interposed between electronic components to prevent coupling between them.

Signal. The radio waves sent out by a transmitter. The electric current induced by radio waves in the antenna. Any alternating current that a vacuum tube circuit is designed to operate from. The input and output voltages or currents in an amplifier.

Spark. An intermittent discharge between electrodes.

Stability. The characteristic of a vacuum tube amplifier circuit that makes its operation reproducible and reliable.

Stage. A circuit, usually an amplifier, which is repeated in cascade to provide higher performance than the circuit alone.

Superheterodyne. A receiver that uses the heterodyne principle to receive high frequency signals and convert them to lower intermediate frequencies for amplification. ("Super" for high frequency combined with "heterodyne".)

Tetrode. A four element tube or screen grid tube.

Tickler. A coil, usually adjustable, that couples a signal from the output of a vacuum amplifier or detector to the input.

Tracking. In a multi-stage receiver the ability for all the tuning capacitors to operate in synchronism so that they can be ganged together for one-knob control.

Transformer, RF and AF. An electric component consisting of two coils or windings coupled closely to each other. An AF transformer is usually wound on an iron core. The RF transformer is wound on an insulated form or "air" core.

TRF. see Tuned Radiofrequency.

Trimmer. A small variable capacitor usually factory adjusted but occasionally provided for adjustment by the operator.

Tube. In general, a glass envelope containing electrodes designed for use in radio circuits. Specifically a triode or other type of vacuum tube for radio use.

Tuned Circuit. A circuit consisting of an inductor and a capacitor that produces a maximum current or voltage when tuned to the incoming signal frequency. See Selectivity.

Tuned Radio Frequency. A generic name for a multistage amplifier consisting of more than one tuned amplifier.

Tuning. The adjustment of a tuned circuit to select a radio station of the desired frequency. See Selectivity, Tuned circuit.

Vacuum Tube. A radio tube having a high vacuum as opposed to tubes having a residual trace of gas.

Variocoupler. A device consisting of a fixed coil and a rotating coil so that the inductive coupling between the two coils can be easily adjusted.

Variometer. A variable inductor. Similar in construction to the variocoupler except that the two coils are connected to form a single inductor. The coupling between coils adds or subtracts to the overall inductance as the movable coil is rotated.

Volume. A control that varies the gain of a receiver and ultimately the volume of sound emitted by the headphones or loud-speaker.

References

Abreviations used in References:

Proc. I.R.E.— Proceedings of the Institute of Radio Engineers

Trans. I.E.E.E.—Transactions of the Institute of Electronic and Electrical Engineers

Archer, Gleason L. 1938. History of Radio to 1926. New York, American Historical Society.

Armstrong, Edwin H. 1922. Some Recent Developments of Regenerative Circuits. Radio News, October, pp. 618-619, 678, 680, 682.

Armstrong, Edwin H. 1924. The Superheterodyne—Its Origin, Development, and Some Recent Improvements. Proc. I.R.E., pp. 539-552.

Batcher, R. R. 1925. The Design of the Grebe Synchrophase. QST, April, pp. 13-16.

Blatterman, A. S. 1926. The Making of a Single-Control Receiver. QST, April, pp. 21-23.

Browning, Glenn H. 1925. The Regenaformer. QST, April, pp. 21-23.

Browning, Glenn H. 1924. A Selective High Mu Receiver. Radio Broadcast, December, pp. 282-287.

Burke, C. T. 1926. Amplifier Ins and Outs. QST, June, pp. 25-28.

Carlton, J. T. 1925. The Counterphase Circuit. Radio News, November, pp. 616-617.

Clement, L. M. 1920. Design of a Radio Receiving Set. Radio News, July, pp. 10-11, 76-77.

Dalton, W. M. 1975. How Radio Began. Vols. 1, 2 and 3. Bristol, Adam Hilger.

Douglas, Alan. 1988. Radio Manufacturers of the 1920's. (In Three Volumes). New York: The Vestal Press.

Dreyer, John F., Jr. and Ray H. Manson. 1926. Proc. I.R.E., April, pp. 217-247, 395-412.

Durkee, Charles H. 1923. 1300 Miles on a One Foot Loop. Radio Broadcast, vol. 2, April, pp. 472-476.

Fitch, Clyde J. 1923. Simple Reflex Circuit. Radio News, July, p. 27.

French, Benedict V. K. 1929. Designing R.F. Circuits for the 224. Radio Broadcast, September, pp. 290-292.

Gherardi, Bancroft and Frank B. Jewett. 1919. Telephone Repeaters. Trans. A.I.E.E., vol. 38, pp. 1287-1345.

Goldsman, J. L. 1923. The Reflex Circuit. Radio News, February, p. 1455.

Goldsmith, A. N. 1926. Reduction of Interference in Broadcast Reception. Proc. I.R.E., vol. 14, pp. 575-603.

Griffin, William J. 1926. New Developments in Radio Apparatus. Radio News, April, pp. 1414-1417, 1497.

Grimes, David H. 1923. Using the "Inverse Duplex" with Various Kinds of Tubes. Radio Broadcast, vol. 3, July, pp. 197-202.

Harrison, Arthur P. 1979. Single-Control Tuning: An Analysis of an Innovation. Technology and Culture, April, pp. 296-321.

Harrison, Arthur P., Jr. 1983. The World versus RCA: Circumventing the Superhet. I.E.E.E. Spectrum, February, pp. 67-77.

Hazeltine, L. A. 1923. The Neutrodyne Receiver. Radio News. May. pp. 1949, 2052.

Hull, A. W. and N. H. Williams. 1924. Physical Review, vol. 23, p. 299.

Hull, L. M. 1925. A True Cascade RF Amplifier. QST, October, pp. 8-11.

Lacault, Robert E. 1924. The Ultradyne Receiver. Radio News, February, pp. 1058-1060.

Laing, A. K. 1926. New Developments in Radio Apparatus. Radio News, February, pp. 1115-1116.

Leutz, Charles R. 1924. Super-Heterodyne Receivers. New York, Experimenters Information Service. (Reprinted 1990 by Frank Kodousek, Milwaukee).

Lewis, W. Turner. 1925. The Radiodyne Receiver. QST, June, pp. 21-22.

Lewis, Tom. 1991. Empire of the Air. New York, Harper-Collins.

Livingstone, Edward A. 1925. The De Forest D17 Receiver. QST, August, pp. 16-19.

Llewellyn, F. B. 1957. The Birth of the Electron Tube. Radio and Television News, March, pp. 43-45.

McLaughlin, J. L. 1924. A One-Control Neutrodyne, "The Super Calamityplex". QST, August, pp. 9-12.

Mesa, J. O. 1927. The Equaphase. Radio Broadcast, November, pp. 42-43.

Minnium, Byron B. 1925. The Isofarad Receiver. Radio News, December, pp. 797, 915.

Minnium, Byron B. 1926. The Improved Isofarad Receiver. Radio News, February, pp. 1152, 1166.

Morecroft, John H. 1927. Principles of Radio Communication. New York: John Wiley & Sons.

Pfaff, Ernest R. 1926. An Improved Laboratory Super-Heterodyne. Radio News, January, pp. 982-983, 1024.

QST. 1923. The Grebe CR-13. December, pp. 28-29.

QST. 1924a. A Neutrodyne-Superheterodyne. June, pp. 12-17.

QST. 1924b. Grebe Developments. October, pp. 36-38.

Radio News. 1923. Notes on the WD-11 Tube. April, p. 1802.

Radio News. 1924. Improvements on the Super-Heterodyne Receiver. May, p. 1576.

Radio News. 1926a. Complete Radio-Frequency Shielding. June, p. 1647.

Radio News. 1926b.A Completely Shielded Neutrodyne. February, p. 1116.

Radio News. 1926c. Harmony in Cabinet Construction. April, p. 1415.

Radio News. 1926d. No Batteries. January, pp. 960-961, 1078

Radio News. 1926e. Single Control on a Distinctive Panel. April, p. 1416.

Radio News. 1926f. A T.R.F. Set of Novel Design. April, pp. 1417, 1497.

Radio News. 1927a. Fewest Possible Controls on New Six-Tube

Set. May, p.1324.

Radio News. 1927b.Loop-Operated Set is Completely Self-Contained. May, p. 1325.

Radio News. 1927c. Rejector Stage Enhances Selectivity of Set. March, pp. 1101-1102.

Radio News. 1927d. Single-Control 6-Tube T.R.F. January, p. 793.

Radio News. 1927e. Skillful Engineering in Seven Tube Receiver. May, p. 1326.

Radio News. 1927f. Successful One-Dial Tuning in Seven Tube Set. April, p. 1213.

Radio News. 1927g. Unique Six-Tube Receiver. January, p. 792.

Radio News. 1927h. Unusual Construction Marks Nine-Tube Receiver. July, p. 19.

Radio News. 1928. Screen-Grid Short-Wave Set Enclosed in Metal Case. December, pp. 538-539.

Rider, John. F. 1928. The A. C. "Bandbox". Radio Broadcast, March, pp. 369-372.

Rowe, G. C. B. 1926a. New Developments in Radio Receivers. Radio News, January, pp. 960-961, 1078-1079.

Rowe, G. C. B. 1926b. New Developments in Radio Apparatus. Radio News, May, pp. 1544-1545.

Rowe, G. C. B. 1926c. New Developments in Radio Apparatus. Radio News, June, pp. 1646-1647.

Silver, McMurdo. 1925a. Revamping the Silver Super-Heterodyne. Radio Broadcast, January, pp. 498-506.

Silver, McMurdo. 1925b. The Super-Autodyne. Radio Broadcast, July, pp. 376-386.

Silver, McMurdo. 1925c. A New Super-Heterodyne. Radio News, October, pp. 444-445, 532.

Snow, H. A. 1924. A Study of Superheterodyne Amplification. QST, October, pp. 20-25.

Stark, Kimball H. 1923. The "Neutrodyne" Receiving System. Radio Broadcast, May, pp. 38-41.

Starkey, Healdon R. 1924. Hassel's Super-Zenith Circuit. QST, November, pp. 28-30.

Stokes, John W. 1982. 70 Years of Radio Tubes and Valves. New York: Vestal Press.

Thompson, Roy E. 1919. The Unitrol Receiver. Proc. I.R.E., vol. 7, pp. 499-516.

Tuska, C. D. 1923. The Superdyne Receiver. QST, November, pp. 7-12.

Tyne, Gerald F.J. 1977. Saga of the Vacuum Tube. Indianapolis: Howard W Sams.

Uehling, E. A. 1929. Design Details of the FADA Set. Radio Broadcast, July, pp. 171-173.

Wheeler, Harold A. and W. A. MacDonald. 1931. Theory and Operation of Tuned Radio-Frequency Amplifier Systems. Proc. I.R.E., May, pp. 738-803.

Radio Manufacturers

The following manufactures are referred to in the text by the shortened form shown in the left column.

A-C Dayton	A-C Electrical Mfg. Co. Dayton, Ohio
Adams-Morgan	Adams-Morgan Company Upper Montclair, N. J.
Appleby	Appleby Manufacturing Co.
Atwater Kent	Atwater Kent Mfg. Co. Philadelphia, Pa.
Bosch	American Bosch Magneto Corp. Springfield, Mass.
Bremer-Tully	Bremer-Tully Mfg. Co. Chicago, Ill.
Clapp-Eastham	Clapp-Eastham Co. Boston, Mass.
Crosley	The Crosley Radio Corp. Cincinnati, Ohio
Day-fan	Day-Fan Electric Co. Dayton, Ohio
De Forest	De Forest Radio Tel. and Tel. Co. New York, N.Y.
Dictograph	Dictograph Products Corp.
ERLA	Electrical Research Laboratories Chicago, Ill.
FADA	F. A. D. Andrea, Inc. New York, N.Y.

Ferguson	J. B. Ferguson, Inc. Long Island City, N.Y.
Freed-Eisemann	Freed-Eisemann Radio Corp. New York, N.Y.
Freshman	Chas. Freshman Co., Inc. New York, N.Y.
GE	General Electric Company Schenectady, Pa.
Golden-Leutz	Golden-Leutz, Inc. New York, N.Y.
Grebe	A. H. Grebe & Co., Inc. Richmond Hill, N.Y.
Hazeltine	Hazeltine Corporation
Jones, Lester	Lester L. Jones / Melco New York, N.Y.
Kennedy	The Colin B. Kennedy Co. San Francisco, Calif.
King	King Quality Products, Inc. Buffalo, N.Y.
Leutz	See Golden-Leutz
Magnavox	The Magnavox Company Oakland, Calif.
Mohawk	All-American Mohawk Corp. Chicago, Illinois
Music Master	Music Master Corp. Philadelphia, Pa.
National	National Co., Inc. Cambridge, Mass.
Perlesz	Perlesz Radio Corp.
Pfanstiel	Pfanstiel Radio Co. Chicago, Ill.
Radio Craft	Radio Craft, Inc. Brooklyn, N.Y.
RCA	Radio Corporation of America New York, N.Y.
Scott	Scott Transformer Co. Chicago, Ill.

Silver-Marshall	Silver-Marshall, Inc. Chicago, Ill.
Sleeper	Sleeper Radio Corp. Long Island City, N.Y.
Sparks-Withington	The Sparks-Withington Co. Jackson, Mich.
Stewart-Warner	Stewart-Warner Speedometer Corp. Chicago, Ill.
Stromberg-Carlson	Stromberg-Carlson Telephone Mfg. Co. Rochester, N.Y.
Thermiodyne	Thermiodyne Radio Corp. Plattsburg, N.Y.
Thompson	R. E. Thompson Mfg. Co. Jersey City, N.Y.
Tuska	The C. D. Tuska Company New York, N.Y.
Walbert	Walbert Mfg. Co.
Westinghouse	Westinghouse Electric and Mfg. Co. East Pittsburg, Pa.
Zenith	Zenith Radio Corp. Chicago, Ill.

Index

S

T

V

W

About the Author

Mr. Rutland obtained his Master's degree in Electrical Engineering from Caltech in 1946. In 1950 he was one of three engineers in charge of the construction of the first digital computer on the West Coast, the National Bureau of Standards, SWAC. This project, using over 3000 vacuum tubes, started him on a career in digital computer hardware. He has served as President and founder of two electronic companies manufacturing computer graphics and image processing hardware and software. His first-hand experience in the design of vacuum tube circuits has been the incentive for the description of 1920's radio circuits described in this book. He is a member of the North West Vintage Radio Society, the Antique Radio Society and the Antique Radio Club of America. His hobbies include vintage radios, wooden ship models and photography. Mr. Rutland is now retired and living with his wife in Oregon.

For extra copies of this book send

$18.95 plus $2.00 postage and handling to:

Wren Publishing
P.O. Box 1084
Philomath, OR 97370

or telephone

(503) 929-4498

Notes

Notes

Notes

Notes

Notes